Reviews of Environmental Contamination and Toxicology

VOLUME 189

Reviews of Environmental Contamination and Toxicology

Continuation of Residue Reviews

Editor
George W. Ware

Associate Editor
David M. Whitacre

VOLUME 189

Springer

Springer
New York: 233 Spring Street, New York, NY 10013, USA
Heidelberg: Postfach 10 52 80, 69042 Heidelberg, Germany

ISSN 0179-5953
ISBN 978-1-4419-2250-2 e-ISBN 978-0-387-35368-5
Printed on acid-free paper.

springer.com

Foreword

International concern in scientific, industrial, and governmental communities over traces of xenobiotics in foods and in both abiotic and biotic environments has justified the present triumvirate of specialized publications in this field: comprehensive reviews, rapidly published research papers and progress reports, and archival documentations. These three international publications are integrated and scheduled to provide the coherency essential for nonduplicative and current progress in a field as dynamic and complex as environmental contamination and toxicology. This series is reserved exclusively for the diversified literature on "toxic" chemicals in our food, our feeds, our homes, recreational and working surroundings, our domestic animals, our wildlife and ourselves. Tremendous efforts worldwide have been mobilized to evaluate the nature, presence, magnitude, fate, and toxicology of the chemicals loosed upon the earth. Among the sequelae of this broad new emphasis is an undeniable need for an articulated set of authoritative publications, where one can find the latest important world literature produced by these emerging areas of science together with documentation of pertinent ancillary legislation.

Research directors and legislative or administrative advisers do not have the time to scan the escalating number of technical publications that may contain articles important to current responsibility. Rather, these individuals need the background provided by detailed reviews and the assurance that the latest information is made available to them, all with minimal literature searching. Similarly, the scientist assigned or attracted to a new problem is required to glean all literature pertinent to the task, to publish new developments or important new experimental details quickly, to inform others of findings that might alter their own efforts, and eventually to publish all his/her supporting data and conclusions for archival purposes.

In the fields of environmental contamination and toxicology, the sum of these concerns and responsibilities is decisively addressed by the uniform, encompassing, and timely publication format of the Springer triumvirate:

Reviews of Environmental Contamination and Toxicology [Vol. 1 through 97 (1962–1986) as Residue Reviews] for detailed review articles concerned with any aspects of chemical contaminants, including pesticides, in the total environment with toxicological considerations and consequences.

Bulletin of Environmental Contamination and Toxicology (Vol. 1 in 1966) for rapid publication of short reports of significant advances and discoveries in the fields of air, soil, water, and food contamination and pollution as well as methodology and other disciplines concerned with the introduction, presence, and effects of toxicants in the total environment.

Archives of Environmental Contamination and Toxicology (Vol. 1 in 1973) for important complete articles emphasizing and describing original experimental or theoretical research work pertaining to the scientific aspects of chemical contaminants in the environment.

Manuscripts for *Reviews* and the *Archives* are in identical formats and are peer reviewed by scientists in the field for adequacy and value; manuscripts for the *Bulletin* are also reviewed, but are published by photo-offset from camera-ready copy to provide the latest results with minimum delay. The individual editors of these three publications comprise the joint Coordinating Board of Editors with referral within the Board of manuscripts submitted to one publication but deemed by major emphasis or length more suitable for one of the others.

Coordinating Board of Editors

Preface

The role of *Reviews* is to publish detailed scientific review articles on all aspects of environmental contamination and associated toxicological consequences. Such articles facilitate the often-complex task of accessing and interpreting cogent scientific data within the confines of one or more closely related research fields.

In the nearly 50 years since *Reviews of Environmental Contamination and Toxicology* (formerly *Residue Reviews*) was first published, the number, scope and complexity of environmental pollution incidents have grown unabated. During this entire period, the emphasis has been on publishing articles that address the presence and toxicity of environmental contaminants. New research is published each year on a myriad of environmental pollution issues facing peoples worldwide. This fact, and the routine discovery and reporting of new environmental contamination cases, creates an increasingly important function for *Reviews*.

The staggering volume of scientific literature demands remedy by which data can be synthesized and made available to readers in an abridged form. *Reviews* addresses this need and provides detailed reviews worldwide to key scientists and science or policy administrators, whether employed by government, universities or the private sector.

There is a panoply of environmental issues and concerns on which many scientists have focused their research in past years. The scope of this list is quite broad, encompassing environmental events globally that affect marine and terrestrial ecosystems; biotic and abiotic environments; impacts on plants, humans and wildlife; and pollutants, both chemical and radioactive; as well as the ravages of environmental disease in virtually all environmental media (soil, water, air). New or enhanced safety and environmental concerns have emerged in the last decade to be added to incidents covered by the media, studied by scientists, and addressed by governmental and private institutions. Among these are events so striking that they are creating a paradigm shift. Two in particular are at the center of ever-increasing media as well as scientific attention: bioterrorism and global warming. Unfortunately, these very worrisome issues are now superimposed on the already extensive list of ongoing environmental challenges.

The ultimate role of publishing scientific research is to enhance understanding of the environment in ways that allow the public to be better informed. The term "informed public" as used by Thomas Jefferson in the age of enlightenment conveyed the thought of soundness and good judgment. In the modern sense, being "well informed" has the narrower meaning of having access to sufficient information. Because the public still gets most of its information on science and technology from TV news and reports, the role for scientists as interpreters and brokers of scientific information to the public will grow rather than diminish.

Environmentalism is the newest global political force, resulting in the emergence of multi-national consortia to control pollution and the evolution of the environmental ethic. Will the new politics of the 21st century involve a consortium of technologists and environmentalists, or a progressive confrontation? These matters are of genuine concern to governmental agencies and legislative bodies around the world.

For those who make the decisions about how our planet is managed, there is an ongoing need for continual surveillance and intelligent controls, to avoid endangering the environment, public health, and wildlife. Ensuring safety-in-use of the many chemicals involved in our highly industrialized culture is a dynamic challenge, for the old, established materials are continually being displaced by newly developed molecules more acceptable to federal and state regulatory agencies, public health officials, and environmentalists.

Reviews publishes synoptic articles designed to treat the presence, fate, and, if possible, the safety of xenobiotics in any segment of the environment. These reviews can either be general or specific, but properly lie in the domains of analytical chemistry and its methodology, biochemistry, human and animal medicine, legislation, pharmacology, physiology, toxicology and regulation. Certain affairs in food technology concerned specifically with pesticide and other food-additive problems may also be appropriate.

Because manuscripts are published in the order in which they are received in final form, it may seem that some important aspects have been neglected at times. However, these apparent omissions are recognized, and pertinent manuscripts are likely in preparation or planned. The field is so very large and the interests in it are so varied that the Editor and the Editorial Board earnestly solicit authors and suggestions of under-represented topics to make this international book series yet more useful and worthwhile.

Justification for the preparation of any review for this book series is that it deals with some aspect of the many real problems arising from the presence of foreign chemicals in our surroundings. Thus, manuscripts may encompass case studies from any country. Food additives, including pesticides, or their metabolites that may persist into human food and animal feeds are within this scope. Additionally, chemical contamination in any manner of air, water, soil, or plant or animal life is within these objectives and their purview.

Manuscripts are often contributed by invitation. However, nominations or new topics or topics in areas that are rapidly advancing are welcome. Preliminary communication with the Editor is recommended before volunteered review manuscripts are submitted.

Tucson, Arizona G.W.W.

Table of Contents

Foreword v
Preface vii

Chemistry and Fate of Simazine 1
AMRITH S. GUNASEKARA, JOHN TROIANO, KEAN S. GOH, AND RONALD S. TJEERDEMA

Ethanol Production: Energy, Economic, and Environmental Losses 25
DAVID PIMENTEL, TAD PATZEK, AND GERALD CECIL

Arsenic Behaviour from Groundwater and Soil to Crops: Impacts on Agriculture and Food Safety 43
ALEX HEIKENS, GOLAM M. PANAULLAH, AND ANDY A. MEHARG

Health Effects of Arsenic, Fluorine, and Selenium from Indoor Burning of Chinese Coal 89
LIU GUIJIAN, ZHENG LIUGEN, NURDAN S. DUZGOREN-AYDIN, GAO LIANFEN, LIU JUNHUA, AND PENG ZICHENG

Mercury Content of Hair in Different Populations Relative to Fish Consumption 107
KRYSTYNA SROGI

Toxicology of 1,3-Butadiene, Chloroprene, and Isoprene 131
HARRELL E. HURST

Index 181

Rev Environ Contam Toxicol 189:1–23

Chemistry and Fate of Simazine

Amrith S. Gunasekara, John Troiano, Kean S. Goh, and
Ronald S. Tjeerdema

Contents

I. Introduction 1
II. Chemistry 2
A. Physicochemical Properties 2
B. Synthesis 4
C. Mode of Action 4
III. Chemodynamics 4
A. Soil 4
B. Water 11
C. Air and Precipitation 13
IV. Degradation 14
A. Abiotic 14
B. Biodegradation 15
Summary 17
Acknowledgments 18
References 18

I. Introduction

Simazine (6-chloro-*N*,*N*′-diethyl-1,3,5-triazine-2,4-diamine) was first introduced in 1956 by the Swiss company J.R. Geigy (Cremlyn 1990). It has been widely used for preemergence control of broadleaf weeds and annual grasses in both agricultural and noncrop fields. For example, in California it was the 28th most used pesticide in 2003 (306,100 kg), with applications primarily on fruit and vegetable crops (CDPR 2003a,b). Simazine is also used as an algicide in fish farm ponds, aquariums, and cooling towers. However, it is toxic to some aquatic animals. For example, Sanders (1969) found that at 22°C 50% lethal concentration (LC_{50}) of simazine for the macrocrustacean amphipod *Gammarus lacustris* was 30, 21, and 13 mg/L

Communicated by Ronald S. Tjeerdema

A.S. Gunasekara (✉) · R.S. Tjeerdema
Department of Environmental Toxicology, College of Agricultural and Environmental Sciences, One Shields Avenue, Meyer Hall, University of California, Davis, CA 95616-8588, U.S.A.

J. Troiano · K.S. Goh
Department of Pesticide Regulation, California Environmental Protection Agency, 1001 I Street, Sacramento, CA 95812-4015, U.S.A.

after 24-, 48-, and 96-hr exposures, respectively. Also, the effect of the herbicide on the water snail (*Lymnea stagnalis*) was such that at a water concentration of 2 mg/L all embryos were destroyed within 9 d; the (50% effective dose) ED_{50} being 0.02 mg/L (Kosanke et al. 1988).

Simazine is available as a powder, liquid, or granular commercial formulation and is the active ingredient (a.i.) in Princep® Caliber 90 (90% a.i.), Princep® Liquid (42% a.i.), Aquazine® (83% a.i.), and other trade name products. Owing to extensive use of the compound as an agricultural herbicide, numerous reports are available regarding its behavior in animals and environmental systems. Although Tennant et al. (2001) have shown that genotoxicity of the compound to mice does not result in any dose-related DNA damage at concentrations as high as 2,000 mg/kg, a number of other reports have shown that it affects several invertebrate species. For instance, at an application rate of 3.4 kg/ha, Edwards (1965) observed reductions in the number of mites, springtails, millipedes, enchytraid worms, and the larvae of *Diptera* and *Coleoptera* for more than 3 mon at a treated site. Ghabbour and Imam (1967) observed that of 10 *Allolobophora caliginosa* earthworms in water, all survived for 14 d at a simazine concentration of 100 mg/L, but they only survived 5 and 2 d at the higher concentrations of 500 and 1,000 mg/L, respectively. Both studies indicate that there is a potential for this herbicide to negatively affect a variety of animals. Additionally, its high use rate and subsequent detection in both surface water and groundwaters have led to concerns about potential impacts on both ecosystem and human health (CDPR 2003a,b; Troiano et al. 2001). Therefore, a review of the chemistry and fate of simazine is useful at this time. Further, to our knowledge there have been no previous reviews published summarizing the environmental fate of simazine.

II. Chemistry

A. Physicochemical Properties

Simazine (Fig. 1) is a member of the triazine family of herbicides; others include prometryn and atrazine (Ware and Whitacre 2004). In pure form it is a colorless white crystalline solid that is thermally stable at temperatures above 150°C and has a density less than that of water (0.43 g/mL; Melnikov 1971). Other physicochemical properties are summarized in Table 1. Of note is that simazine water solubility is reported for four different temperatures. Curren and King (2001) found that its solubility increased more than 10 fold, from 17 to 240 μg/mL, as the temperature was raised from 50° to 100°C. In general, at ambient conditions (20°C and 1 atm) simazine is slightly soluble in water, relatively nonvolatile, capable of moderately partitioning to organic phases, and susceptible to photolysis (Table 1).

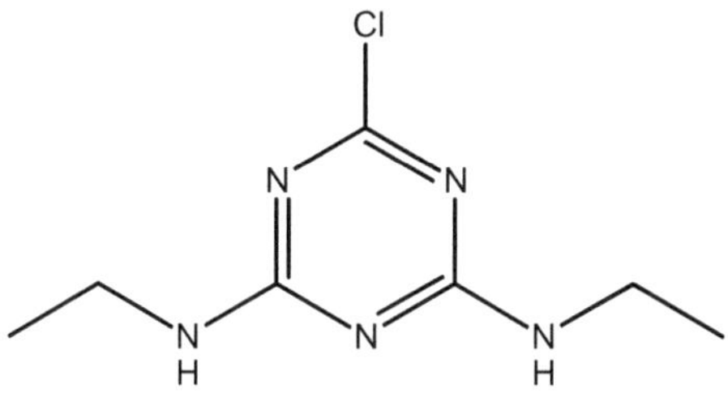

Fig. 1. Chemical structure of simazine.

Table 1. Physicochemical properties of simazine.

Chemical Abstracts Service registry number (CAS #)[a]	122-34-9
Molecular formula[a]	$C_7H_{12}ClN_5$
Molecular weight (g/mol)[a]	201.66
Density at 20°C (g/mL)[a]	0.436
Melting point in (°C)[a]	225–227
Octanol–water partition coefficient (K_{ow})[a]	122
Organic carbon normalized partition coefficient (K_{oc})[d]	130
Aerobic microbial half life ($t_{1/2}$) in sandy loam (d)[a]	91
Anaerobic microbial $t_{1/2}$ in sandy loam (d)[a]	70–77
Photolytic $t_{1/2}$ of 6.7 μg/cm^3 at $\lambda = 53.25$ nm (d)[c]	4.5
Photolytic $t_{1/2}$ in sandy loam under natural light (d)[a]	21
Henry's law constant (atm-m^3/mole) [a]	9.48×10^{-10}
Dissociation constant (pK_a) at 21°C [c]	1.70
Water solubility (mg/L)	
0°C[a]	2.0
20°C[b]	5.0
22°C[a]	6.2
85°C[a]	84
Vapor pressure (mm Hg)	
10°C[a]	9.0×10^{-10}
20°C[a]	6.1×10^{-9}
25°C[a]	2.2×10^{-8}
30°C[a]	3.6×10^{-8}
75°C[a]	4.5×10^{-5}
100°C[a]	9.8×10^{-4}

All parameters are at 25°C unless otherwise specified.
[a]Vencill et al. (2002).
[b]Verschueren (1984).
[c]Montgomery (1997).
[d]Average from Wauchope et al. (1992).

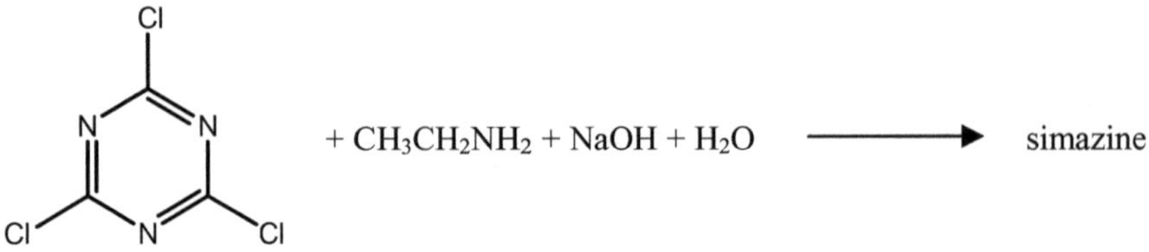

Fig. 2. Synthesis of simazine using cyanuric chloride, ethylamine, and sodium hydroxide in aqueous medium (Melnikov 1971).

B. Synthesis

Simazine can be produced in >90% yield by the reaction of cyanuric chloride with ethylamine and sodium hydroxide in aqueous solution (Fig. 2; Melnikov 1971). It is often sold as a powder in which half the mixture consists of clay (kaolin) or chalk ($CaCO_3$) as diluents (Melnikov 1971). The compound can be hydrolytically dechlorinated when heated with caustic alkali under laboratory conditions (Melnikov 1971).

C. Mode of Action

Simazine inhibits the photosynthetic electron transport process in annual grasses and broadleaf weeds (Ware and Whitacre 2004). Wilson et al. (1999) provides a detailed characterization of its mechanism of inhibition. According to Cremlyn (1990), its uptake is via the roots of emerging seedlings. Subsequently, the compound hinders photosynthesis in leaves, causing them to yellow and die (Ware and Whitacre 2004). In contrast, many varieties of maize and sugar cane are resistant to simazine because they possess a specific hydrolase that catalyzes detoxification of the compound (Cremlyn 1990).

III. Chemodynamics

A. Soil

A major concern with many agricultural pesticides is the potential to leach into surface and groundwater systems that are used as drinking water sources. The low sorption capacity of simazine in soil plays an important role in its movement to water systems, given its moderate K_{oc} and K_{ow} values, low water solubility, and low volatilization (Tables 1, 2). A number of reports have employed a variety of soils to characterize the sorption–desorption behavior of the herbicide, as summarized next.

Sorption to Minerals. Sorption of simazine to clays is weak, leading to low soil–water distribution coefficients (K_d). For instance, Cox et al. (2000a,b) reported low K_d values (approximately 13) for the compound's sorption to

Table 2. The presence of simazine in surface water and groundwater in California.

Parameter	Surface water (concentrations in μg/L)			Groundwater (concentrations in μg/L)		
Year	2000	2001	2002	2001	2002	2003
Total number of sites	221	460	147	1019	1173	2347
Number of detections	36	166	4	3	87	17
Maximum concentration	2.892	3.700	0.156	0.252	0.244	0.103
Minimum concentration	0.050	0.011	0.050	0.193	0.034	0.050
Median concentration	0.160	0.024	0.068	0.223	0.104	0.098

All data are from the California Department of Pesticide Regulation Surface and Ground Water Databases, 2004.

montmorillonite clay minerals. The mechanism was attributed to the presence of protonated species of the herbicide, which bound to hydrophobic microsites on the clay surfaces. In addition, simazine was demonstrated to weakly adsorb to both ferrihydrite, an iron oxide mineral, and sand (Table 3; Celis et al. 1997). Low K_d values were also obtained by Oliveira et al. (2001) for five Brazilian sandy soils with low organic matter. When standardized to the organic carbon content of soil they were below 116, which correlated well to other studies also reporting K_{oc} values: 103–152 (Ahrens 1994), 105 (Hassink et al. 1994), and 130 (Flury 1996). Reddy et al. (1992) reported that in mineral-rich organic matter and clay-poor soils simazine does not persist (K_d, 0.33–0.76 mL/g) but readily leaches. They estimated that rainfall of 20–23 cm could displace >42% of applied simazine in sandy soils low in organic matter and clay. This estimation was confirmed by Ritter et al. (1996), who observed transport of simazine to shallow groundwater in sandy soils when more than 30 mm rainfall occurred shortly after application to the soil surface.

Several desorption studies involving mineral soils have also been conducted and confirm the weak sorption of simazine to mineral soils. It readily desorbs from montmorillonite surfaces, indicating weak van der Waals or hydrogen-bonding interactions between the herbicide and mineral sites (Celis et al. 1997, 1998). Similarly, in sandy soils Cox et al. (2000a) reported rapid desorption with no hysteresis. These studies indicate simazine can readily leach in mineral-rich or sandy soils.

Investigations into competitive sorption also support weak sorptive processes for simazine with mineral soils. For instance, Sannino et al. (1999a) reported that sorption decreased with increasing occupation of the

Table 3. Simazine sorption to different sorbates.

Author	Sorbent	Conditions	K_d (mL/g)	K_{oc}	Method
Sannino et al. (1999b)	Montmorillonite	pH = 3.7	458	—	Isotherm
		pH = 5.6	16	—	Isotherm
		pH 3.7 and Al(OH)x at 18 mEq Al/g clay	19	—	Isotherm
Celis et al. (1997)	Montmorillonite	pH = 5.5	13.4	—	Isotherm
Celis et al. (1998)	Montmorillonite		17.0	—	
Reddy et al. (1992)	Fine sand	% OC = 0.50	0.29	58	Isotherm
	Fine sandy loam	% OC = 1.39	1.06	76	
Barriuso et al. (1997)	Soil		0.78	74	Breakthrough curves
	Compost		10.5	62	
Cox et al. (2000b)	Sandy soil		2.9	445	Isotherm
	Sandy soil with OOMW		16	1550	
	Montmorillonite		12.9	1700	
	Montmorillonite with OOMW		21.5	1537	
Cruz-Guzmán et al. (2004)	Montmorillonite (untreated)		28		Isotherm
	Montmorillonite with L-carnitine		50,000		
Brereton et al. (1999)	Black fly silk		19,000		Isotherm

OC, organic carbon; OOMW, organic olive-mill waste.

mineral surface with Al(OH) species (see Table 3). Water molecules can also compete with the herbicide for sorption sites on clay surfaces (Laird et al. 1992). Simazine sorption to mineral soils has been found to be modified by pH. Sannino et al. (1999a) found that large amounts of the compound are adsorbed onto montmorillonite surfaces in a pH-dependent manner, where the K_d value was 28 times greater at pH 3.7 ($K_d = 458$) than at pH 5.6 ($K_d = 16$). The pH-dependent, strong electrostatic sorption was attributed to cation-exchange mechanisms because simazine becomes cationic at low pH, reacting strongly with negative charges on the mineral surface. These studies, and that of Celis et al. (1997), confirm that montmorillonite is the primary soil mineral that contributes to simazine adsorption. In summary, simazine adsorbs to the mineral components of soil, but to a minor extent, and the mineral portion alone lacks the ability to retain residues and restrict its movement through soils (see Table 3).

Sorption to Organic Matter. Partitioning of simazine (see Table 1) to organic matter is not very significant (K_{ow} and K_{oc} ~ 125) in comparison to chemicals bearing a strong affinity for organic matrices, such as DDT (K_{oc} ~ 160,000). In fact, the coefficients are not much greater than the sorption values previously discussed for mineral soils. A number of notable studies have examined the contribution of organic matter (OM) to simazine sorption by soils. Beltran et al. (1998) evaluated its adsorption and desorption by sandy Western Australian soils (see Table 3) and found a positive correlation between simazine retention and soil OM content. However, the equilibrium process was observed to be slow. A water flow rate of 3 m/d through soils containing the herbicide resulted in a lack of equilibrium, as observed by reduced (40%–60%) K_d values, when compared to a static system. The observed K_{oc} values (833 with 6% OM) are still relatively low in terms of its sorption when compared to other more hydrophobic compounds such as DDT.

Reddy et al. (1992) also examined the interactions of simazine with OM by measuring its percent sorption in sandy soils having either 0.50% or 1.39% organic carbon (OC; ~50% of OM). With increasing OM content, sorption increased significantly, from 19% to 46%, for the 0.50% and 1.39% OC-containing sandy soils, respectively (Table 3). The role of OM was further illustrated by Barriuso et al. (1997), who compared sorption to OM-rich and poor materials such as compost and a mineral soil, respectively (Table 3). Sorption to compost (K_d = 10.5 L/kg) was approximately 13 times higher than to soil (K_d = 0.78 L/kg) and was attributed to the greater OM content of compost (16%). Snail pedal mucus and black fly silk can also contribute significantly to the overall sorption of simazine by soils, as Brereton et al. (1999) found sorption to these substances was orders of magnitude greater than to soils alone (e.g., K_d for black fly silk was 19,000; for soils, it was 135; see Table 3).

Sorption of simazine to OM has been shown to both increase its sequestration and delay its degradation. Laabs et al. (2002) used ^{14}C-labeled simazine to compare the degradation rate and bound residue formation between an Ustox soil rich in OM and a Psamment tropical soil lacking OM. When applied at a rate of 2 kg/ha, simazine persisted more in the organic-rich soil; the 50% dissipation time (DT_{50}) was 27 d compared to 14 d in the sandy Psamment soil. The slower dissipation was due to a greater nonextractable portion of residue, which was at 55%–60% of the applied simazine. In another analysis, Cox et al. (2001) found that different organic amendments had an influence on reducing herbicide movement in soils. For example, they found that organic amendments to a mineral soil greatly enhanced its sorption, thereby reducing potential leaching. Furthermore, solid organic amendments provided greater adsorption than liquid forms. Overall, sorption of the herbicide in soil amended with solid OM was greater by a factor of 2.5 compared to nonamended soil. Additionally, the study found that soil amended with liquid forms of OM competed with the herbicide for sorption sites. Similarly, Celis et al. (1998) found that when soil minerals (ferrihydrite and montmorillonite) are in association with OM fractions, such as humic acid, the sorption capacity for simazine is greatly enhanced.

Ertunç et al. (2002) reported a fast rate of simazine sorption to compost. Aqueous extractions after 29 and 200 d composting, equivalent to thermophilic and mesophilic phases, respectively, produced only 4.2% and 3.1% of the herbicide, respectively. Such data coincide well with the current hypothesis that organic substances in aged soils have a greater affinity and "holding-capacity" for pesticides (Garcia-Valcarcel and Tadeo 1999). The nonextractable fraction of the herbicide was greater (64%) after 29 d composting. Also observed was a distinct shift from rather weak interactions to strong covalent linkages for simazine and its major metabolites (Fig. 3) as composting time increased.

Other studies have focused on understanding the molecular-level binding properties of the herbicide with OM. Cox et al. (2001) suggested a partition mechanism. However, according to Celis et al. (1998) the main mechanism between simazine and OM, primarily humic substances, appears to involve hydrogen bonding and proton transfer processes. More recent

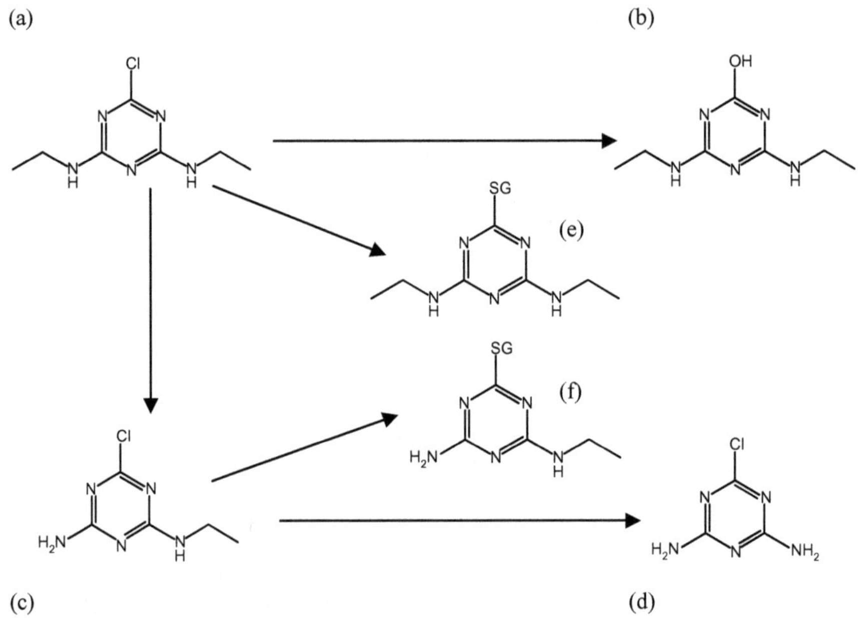

Fig. 3. The general abiotic and biotic degradation pathways of simazine (*a*) where 2-hydroxy-4,6-bis(ethlyamino)-*s*-triazine (*b*) is produced by hydrolysis. Photolytic and biotic oxidation of simazine produce deethyl simazine (*c*) and diamino chlorotriazine (*d*) (Evgenidou and Fytianos 2002), while phase II metabolism in higher-order organisms leads to glutathione conjugates (*e*) and (*f*) from simazine and deethyl simazine, respectively (Adams et al. 1990).

work has attributed the binding affinity to the presence of functional groups, such as carboxylic acids, that are readily available in organic substances. For example, Cruz-Guzmán et al. (2004) examined the sorption of simazine to functional group-rich organics inserted within the interlayers of montmorillonite. The results showed an unprecedented 1,000- to 2,500-fold increase in the sorption capacity (K_f) or Freundlich coefficients [units = (μg/g)/(μg/mL)^N]; the untreated montmorillonite K_f was 28, compared to 50,000 for L-carnitine-containing montmorillonite (see Table 3).

A significant sorptive component of OM is dissolved organic matter (DOM), defined as the fraction of OM than can pass through a 0.45-μm membrane filter. Cox et al. (1999) measured greater simazine sorption to a sandy soil when it was amended with a 26% liquid organic solution; sorption was increased by a factor of 2.5 when amended to 20% w/w and by a factor of 1.8 when amended to 10% w/w. Subsequent studies (Cox et al. 2000b) found that in a sandy soil a good sorbent for the herbicide was well-humified solid organic olive-mill waste, which increased the isotherm sorption capacity of the soil (see Table 3). Simazine distribution in compost DOM was mainly associated with the low molecular DOM fraction (60% simazine associated with <1 kDa fraction) for up to 200 d composting (Hartlieb et al. 2001). This study also found that 16% of the initial concentration (1.17 mg/kg) was mineralized after 370 d, and another 65% was nonextractible and bound to the compost matrix.

Although low molecular weight fractions of OM can act as a major sorbent for simazine, Matsui et al. (2002) observed that they also compete for sorption sites on high-affinity water remediation materials such as activated carbon. Using microcolumn experiments, Matsui et al. (2002) reported that low molecular weight molecules competed directly with the herbicide for 7–12 Å micropore sites on activated carbon. Cox et al. (2000b) found that DOM produced as a result of a liquid organic amendment will compete with simazine molecules for sorption sites on montmorillonite and increase the potential for simazine to leach from the soil; its sorption to the organic-amended clay had approximately the same K_{oc} as the nonamended clay (see Table 3). Thus, DOM not only can enhance the sorption of simazine in soil but in mineral soils may compete with the compound for sorption sites.

In a practical application of these findings, Davis and Lydy (2002) conducted a 3-yr runoff study at a golf course incorporating enhancements in filtration. After spring and summer applications, rainfall events caused simazine to move at high concentrations into two centrally located ponds. The presence of simazine caused significant reductions in both macroinvertebrate diversity and population size. These changes were remediated upon incorporation of best management practices around the ponds, which consisted of establishing buffer zones of aquatic vegetation and rerouting drainage systems to a maintained filtration area. The herbicide was eliminated from the two ponds along with complete recovery of the diversity and population size of the macroinvertebrates in all ponds.

Soil Moisture Influence. Soil moisture has been shown by many studies to contribute to the sorption and dissipation of simazine. Wang et al. (1996) found a positive correlation between soil moisture content and degradation. Garcia-Valcarcel and Tadeo (1999) examined the effects of 10% or 18% moisture content and 4%–18% wet–dry cycles on the degradation rate of simazine in a sandy loam soil. K_f averages of 0.9, 0.848, and 1.215 corresponded to 10%, 18%, and 4%–18% soil moisture content, respectively. Shrinkage and swelling of the soil matrix as a result of wet–dry cycles produced the highest K_f [ranging from 0.5 to 1.2 (μg/g)/(μg/mL)^N]. The $t_{1/2}$ values ranged from 27 to 126d, depending on the soil moisture content. A similar $t_{1/2}$ value was obtained by Wang et al. (1996), 34d with 20% field capacity. Louchart et al. (2001) found that two phases governed the breakdown of simazine in moisture-influenced soil conditions of southern France. The first phase was rapid degradation of the compound and was attributed to soil moisture, which promoted microbial activity. The second phase was slow because of the dry conditions and decreased microbial activity, typical of the Mediterranean climate.

Influence of Agricultural Management Practices. Farming practices can influence simazine movement in agricultural soils. For instance, Louchart et al. (2001) compared runoff in no-till and tilled fields in southern France under Mediterranean climatic conditions and found that tilled fields had lower runoff concentrations as a result of increased infiltration. In agreement, Lennartz et al. (1997) reported that in fields that had not been tilled there was no vertical transport below a soil depth of 2cm (18%, 55%, and 26% clay, silt, and sand, respectively) under the same climatic conditions as described in Louchart et al. (2001). However, in tilled sites Lennartz et al. (1997) found the herbicide extending to depths of 15cm in a soil of 22% clay, 48% silt, and 29% sand. These findings are supported by Glenn and Angle (1987), who found that no-till fields had greatly increased residue runoff. They also found that conventional tillage reduced simazine in runoff water by 70% as compared with no-till or conservation tillage corn plots. However, tillage seems to also reduce dissipation time of simazine in the soil: 85 and 281.5d for DT_{50} and DT_{90}, respectively, in no-till fields, and 157 and 523d for DT_{50} and DT_{90}, respectively, in tilled fields (Lennartz et al. 1997). Similarly, Cogger et al. (1998) found $t_{1/2}$ values to be 175 and 424d for the first and second year of herbicide application, respectively, and it remained in the soil up to 4yr after application. These studies suggest that tillage reduces simazine runoff and dissipation time in agricultural soils.

In orchards, the areas between tree furrows often are not tilled. Because simazine is extensively used in orchards, its runoff into surface waters could potentially be high. Troiano and Garretson (1998) examined its movement in runoff water from citrus orchard row middles (the areas between furrows), as affected by mechanical incorporation, using a Handford Sandy Loam. Simazine was applied to the plots at a rate of 2.2kga.i./ha. In the first simulated rainfall event (water delivery rate, 22 mm/hr), the herbicide collected in runoff water was only 10% in plots where it was mechanically incorporated compared to runoff collected from areas where the row middles were left undisturbed. Mechanical incorporation of the row middles drastically reduced its concentration in runoff water but also reduced the amount of water that was collected as runoff. The results indicate that orchard row middles with compacted, nontilled soils significantly contribute to simazine concentrations in runoff water. Similarly, Liu and O'Connell (2002) examined the herbicide runoff (application rate, 2 kg/ha a.i.) from a California citrus orchard. Simulated runoff events with 3.5 cm water applied by rotor sprinklers revealed that the nonirrigated, compacted land areas had the largest runoff (6.5% of initial treatment). In a subsequent study, Liu and O'Connell (2003) showed that lower application rates (1 kg/ha a.i.) achieved the same amount of weed control while reducing herbicide loss.

The previous studies were conducted on soils with minimal slope (1%–2%). In tilled red maple nursery plots having 5% slopes, while residues were detected at depths of 60–90 cm after the first rainfall, approximately 0.1%–6.8% of the herbicide was removed in runoff water (Stearman et al. 1997). Cover crops of rye grass and crimson clover were found to significantly reduce the overall runoff concentration of simazine when compared to bare plots (Stearman et al. 1997). However, Ritter et al. (1996) reported higher concentrations in groundwater under tillage conditions. Concentrations on the tilled plots ranged from <0.10 to 16 μg/L and <0.10 to 6 μg/L at 3- and 4.5-m depths, respectively. The authors concluded that the compound could move to shallow groundwater, particularly in the sandy soils of the Mid-Atlantic States, if more than 30 mm rainfall were to occur shortly after application. Such findings are supported by Cogger et al. (1998), who studied the long-term movement of the herbicide through soils with low water tables (<3 m) in western Washington. An application rate of 4.5 kg/ha to raspberry fields with sandy/silt loam (alluvium) soils for 3 yr showed that most of the herbicide was retained in the top 15 cm (between 400 and 2,310 μg/kg), but small amounts, to 15 μg/kg, migrated further into the soil profile (120–180 cm depth). Also, its concentrations in well water samples were related to annual rainfall (10.35 cm). The short-circuiting of water through soil macropores from rainfall or irrigation and kinetic sorption–desorption processes was found to drive its leaching. These studies indicate that mechanical incorporation into agricultural fields could reduce herbicide runoff but, depending on soil texture, residues can also move vertically through the soil.

B. Water

Simazine as a contaminant in water is of concern because of the universal importance of groundwater and surface waters for drinking purposes. Its detection in groundwater was reported as early as 1989 in the Netherlands

(Lagas et al. 1989). Both parent and breakdown products have the potential to move to either surface water or groundwater as indicated by the summary of monitoring conducted in California (see Table 2; Bacey et al. 2004; Spurlock et al. 2000; Troiano et al. 2001). Total concentrations incorporating both parent and breakdown products have been found to exceed health levels established for simazine (Troiano and Nordmark 2002).

Groundwater. Spurlock et al. (2000) examined the distribution of simazine and its by-products, deethyl simazine (see Fig. 3c) and diamino chlorotriazine (Fig. 3d), in 30 domestic water wells in Fresno and Tulare Counties, California. By-products significantly contributed to the overall triazine concentrations measured in groundwater, accounting for 24%–100% of total residues. At least one well had a simazine concentration (3.8 μg/L) approaching the U.S. Environmental Protection Agency (EPA) maximum contaminant level (MCL) of 4 μg/L. A positive correlation was found between simazine and by-product concentrations, although overall herbicide concentrations did not change over the 2-yr sampling. Similar results were obtained by Troiano and Nordmark (2002), where many sampled wells had triazine detections: 110 contained deethyl simazine, 105 had diamino-chlorotriazine, and 85 had simazine. In North Carolina, analysis of 97 groundwater wells revealed the presence of the compound in 5 of the wells at concentrations ranging from 9% to 211% of the state groundwater quality standard (3.5 μg/L; Wade et al. 1998). Other reports confirm that simazine has previously been found in 89 of 1,430 groundwater wells sampled throughout the United States (Holden et al. 1992), including 56 wells in Illinois (Long 1989) and 10 of 20 wells in Wisconsin (Habecker 1989). In Denmark, low herbicide concentrations (<0.1 μg/L) were reported for 2 shallow groundwater wells (Spliid and Koppen 1998). It was also reported that in sandy soil areas, 12 of 184 shallow groundwater wells were contaminated with the herbicide, but at low concentrations (<0.1 μg/L). These studies prove that although simazine is minimally soluble in water, it can be present in groundwater at significant concentrations.

Surface Water. Battaglin and Goolsby (1999) investigated the presence of simazine in surface waters close to agricultural areas of the U.S. and found elevated concentrations in 53 Midwestern rivers. Furthermore, concentrations were related to runoff events that took place 1–3 mon following application. Similarly, Power et al. (1999) examined simazine runoff into the Thames Estuary in the United Kingdom and found that concentrations peaked in the month of June (0.12 and 0.167 μg/L in 1988 and 1997, respectively) and declined in fall. However, they also found that there was approximately a 91% overall reduction in herbicide concentration in the estuary from 1988 to 1997. Battaglin and Goolsby (1999) reported that improved agricultural practices in the U.S., such as split herbicide

Table 4. Water quality criteria (in µg/L).

U.S. EPA MCL (maximum contaminant level)[a]	4
U.S. EPA MCLG (maximum contaminant level goals)[a]	4
North Carolina Ground Water Quality Standard (1998)[b]	3.5

[a]U.S. EPA (2003).
[b]Wade et al. (1998).

applications, decreased per-acre application rates and that increased use of postemergence herbicides reduced the presence of simazine over time in midwestern rivers (1994–1995 compared to 1989–1990).

Another U.S. estuary receiving large amounts of pesticide runoff is the Chesapeake Bay. Simazine usage in the Patuxent River watershed (230 ha) was 1,600 kg a.i. in 1996, and concentrations as high as 0.8 µg/L were reported in the estuary (Harman-Fetcho et al. 1999); the upstream concentration averaged 0.55 µg/L whereas downstream it was 0.04 µg/L. Studies clearly show that simazine loss from agricultural application can be significant; over time it will move downstream and from surface waters into estuaries.

The parent compound and its degradation products (deethyl simazine and diamino chlorotriazine) have also been extensively monitored in California by the Department of Pesticide Regulation. Table 2 summarizes detection of simazine in California surface water and groundwater; in some cases, it also approaches the U.S. MCL of 4 µg/L (Table 4).

C. Air and Precipitation

The potential for simazine to volatize from soils and water is low, as confirmed by the low vapor pressure and Henry's constant values presented in Table 1. However, residues in the atmosphere have been measured, as well as correlated to application time in the field. For instance, Chevreuil et al. (1996) found that atmospheric concentrations peaked after local agricultural application in the spring (June and July 1993). The vertical flux of simazine was also measured by air-sampling and aerodynamic techniques over 24 d after its application (1.68 kg/ha) on a fallow soil by Glotfelty et al. (1989). Calculated volatilization losses in the first 21 d were 0.021 kg/ha (1.3% loss). Daily losses were found to be dependent on moisture content because simazine, applied as a wettable powder, became susceptible to wind erosion as the soil surface dried in the afternoon sun.

Reported concentrations of simazine in precipitation are low when compared to those in surface water and groundwaters. For instance, in 1991 Shertzer et al. (1998) found low concentrations (0.15 µg/L) in rainwater near the Conodoguinet Creek watershed in south-central Pennsylvania, while Albanis et al. (1998) measured a concentration of 0.005 µg/L from rainfall in Greece. Likewise in Paris, France, Chevreuil et al. (1996) reported concentrations between 5 and 650ng/L. The presence of simazine in nonagricultural areas supports the findings that it can be transported in the atmosphere and deposited elsewhere by precipitation. However, according to Wauchope (1978), volatilization losses are much smaller than those via degradation but similar to reported surface runoff concentrations.

IV. Degradation

A. Abiotic

Table 1 presents the physicochemical properties of simazine that characterize its abiotic degradation, including rates of hydrolysis, photolysis, reactivity, and remediation. The major abiotic degradation products are 2-hydroxy-4,6-bis(ethylamino)-*s*-triazine (Fig. 3b), 2-chloro-4-amino-6-(ethylamino)-*s*-triazine (deethyl simazine; Fig. 3c), and 2-chloro-4,6-diamino-*s*-triazine (diamino chlorotriazine; Fig. 3d).

Oxidation. Simazine can be oxidized by ozone through the use of catalytic amounts of Mn^{2+} (0.1–1 mg/L) or Fe^{2+} (1–5 mg/L) according to Rivas et al. (2001). They also found that after 30min oxidation increased from 80% to 90% with the addition of Mn^{2+} (up to 0.2 mg/L). However, use of Fe or Mn at concentrations higher than 1 mg/L inhibited simazine oxidation. Oxidation rate was also observed to increase with pH from 5 to 9 (Rivas et al. 2001). The positive correlation was attributed to the presence of highly reactive hydroxyl radicals, with the main oxidation product [detected by high pressure liquid chromatography (HPLC)] being deethyl simazine (Fig. 3c; Rivas et al. 2001).

Hydrolysis. Simazine hydrolysis was evaluated by Comber (1999), who found that no appreciable degradation occurred over a period of 100d at 15°C in the dark at either pH 7 or pH 9. Also, the presence of iron and aluminum hydroxides did not promote reaction. However, a $t_{1/2}$ of 145d was calculated for its degradation in the dark at pH 4.0 (the lowest pH routinely found in natural waters), indicating that even in favorable conditions hydrolysis occurs at a slow rate. Hydrolysis under low pH conditions leads to the formation of 2-hydroxy-4,6-bis(ethlyamino)-*s*-triazine (Fig. 3b), which does not possess herbicidal properties.

Acid catalysis was examined by Sawunyama and Bailey (2002), who calculated the enthalpy ($\Delta_h H$), Gibbs free energy ($\Delta_h G$), and entropy of hydrolysis ($\Delta_h S$) at 25°C to be −70, −63, and −23 kJ/mol, respectively. Theoretically, such values indicate that simazine hydrolysis is energetically spontaneous, but this was not observed in experimental studies. However, its hydrolysis does appear to be kinetically controlled (Sawunyama and Bailey 2002).

Simazine hydrolysis, as influenced by DOM in natural waters, temperature, and mixing, was examined by Noblet et al. (1996). They found that it exhibited no loss in concentration after 43 d. Likewise, the presence of 70 mg/L DOM had no effect on herbicide transformation at pH 8.0 and 40°C, leading Noblet et al. (1996) to suggest that DOM alone does not influence its transformation or dissipation.

Photolysis. Among the abiotic reactions important for simazine degradation, photochemical reactions play a major role. The $t_{1/2}$ for photodegradation ranges from 4.5 to 21 d (see Table 1), depending on the medium (Vencill et al. 2002; Montgomery 1997). Although in the dark it is not degraded, more than 90% is degraded under UV radiation ($\lambda = 290$ nm) after 25 hr (Evgenidou and Fytianos 2002). Degradation follows pseudo-first-order kinetics, and the $t_{1/2}$ was calculated to be between 2.7 and 5.4 hr, depending on the water source, dissolved oxygen, and OM content. Indirect photolysis, involving UV-generated hydroxyl radicals from hydrogen peroxide, significantly enhanced the simazine degradation rate by up to four times (Evgenidou and Fytianos 2002). Two pathways for simazine photodegradation have been proposed: one involving ring dechlorination and subsequent hydroxyl substitution (Fig. 3b), and the other involving oxidation of the propyl side chains (Evgenidou and Fytianos 2002).

Simazine is also effectively degraded by the photo-assisted Fenton reaction, which also generates hydroxyl radicals. Huston and Pignatello (1999) found that >99% of the initial concentration of the herbicide (0.342×10^{-4} *M*) degraded within 10 min (at 25°C, pH 2.8) when subjected to the following: 0.00005 *M* Fe^{3+}, 0.01 *M* H_2O_2, and 1.2×10^{19} quanta/L/s of fluorescent blacklight UV irradiation (300–400 nm). This reaction was optimal for simazine at pH 2.8, where half of the iron is present as Fe^{3+} and half as the active species $Fe(OH)_2$. However, this pH is rarely encountered in the environment.

B. Biodegradation

The degradation half-life values in Table 1 indicate that simazine could be somewhat persistent in the environment, up to 8 mon. Barriuso et al. (1997) found that herbicide decomposition was similar to atrazine in compost, via triazine ring degradation. Biotic degradation appears to be the most effective means of dissipation when compared to abiotic processes. Strong and coworkers (2002) have isolated a gram-positive bacterial strain of *Arthrobacter aurescens* capable of consuming 3,000 mg/L atrazine in liquid as a sole carbon and nitrogen source, via catabolism, to supplement its growth; it was also capable of degrading simazine. Similarly, Martin-Montalvo et al. (1997) found an aerobic bacterium strain (DSZ1) capable of growing solely on simazine at concentrations of 5, 10, and 20 mg/L; growth takes place with a lag phase. Approximately 70% of the herbicide at a concentration of 5 mg/L was degraded in 19 d at optimal conditions with a density of 10^6 cells/mL. Cyanuric acid was successfully used as a substrate, and the authors suggest it could be an intermediate in the simazine catabolic degradation pathway (Martin-Montalvo et al. 1997). In a similar study, a bacterial strain of *Moraxella ovis* was capable of completely degrading 200 mg/L of the herbicide within 5 d (Kodama et al. 2001). This bacterium grows well at high pH, but values around 5 are most suitable for simazine degradation. Cook and Hutter (1984) found that *Rhodococcus corallinus* was capable of transforming deethyl simazine (see Fig. 3c) by dehalogenation, dealkylation, and hydroxylation, but this particular bacterium could not cleave the triazine ring.

Vink and van der Zee (1997a) reported the microbial reductive transformation of simazine to deethyl simazine (Fig. 3c) and diamino chlorotriazine (Fig. 3d) in anaerobic lake sediment with decreasing oxygen concentration (over 200 d). Transformation (>90%) in aerobic and anoxic microbial soil columns was approximately 22 and 53 d, respectively. However, >30% of its initial concentration (4.5 μg/g) remained after 200 d incubation. The microbial transformation of the herbicide in surface waters was also studied by Vink and van der Zee (1997b), and its $t_{1/2}$ was calculated to range between 1 and 139 d. Principal-component analysis (PCA) revealed that three discriminating environmental variables explained 84% of the total variance in the data. The first principal component contained variables that promoted biorespiratory processes, in which a relationship exists between sorption potential, nitrogen, and microbial activity. The second and third components were influenced by macro/micronutrients and phosphorous groups, respectively.

Fungal strains capable of degrading simazine have also been identified. Kodama et al. (2001) identified strain DS6F (*Penicillium steckii*) as capable of degrading 50 mg/L of the herbicide, utilizing it as its sole carbon and nitrogen source. Rate of degradation was improved when easily accessible carbon sources were added into the medium; an increase of 53% was obtained after 5 d cultivation at 30°C with glucose. In contrast, *Phanerochaete chrysosporium*, a common white rot fungus, was not as effective in degrading ^{14}C-labeled simazine (5.4%) into deethyl simazine (Mougin et al. 1997). Analysis of the primary modes of degradation – extracellular lignin and manganese-dependent peroxidases to carry out the N-dealklylation of simazine – proved unsuccessful (Mougin et al. 1994). However, Filazzola et al. (1999) have shown that laccase from *Cerrena unicolor* catalyzes the oxidation of many aromatic compounds to produce polymeric products. Sannino et al. (1999b) found that simazine can be oxidized by laccase from the same fungal strain used by Filazzola et al. (1999). Even more interesting is the finding that simazine inhibited the oxidation of 2,4-dichlorophenoxyacetic acid (2,4-D) by laccase. These results raise important concepts that must be considered when attempting to use bioremediation of polluted sites containing more than one contaminant.

Simazine metabolism in higher-order organisms has also been reported. Hepatic microsomes from rats, mice, goats, sheep, pigs, rabbits, and chickens exposed to simazine revealed the only phase I metabolite is deethyl simazine (see Fig. 3c), an oxidation product mediated by cytochrome P-450 isozymes (Adams et al. 1990). No products resulting from dechlorination or ring cleavage were detected in the phase I studies. Fenton reagent experiments, which generate hydroxyl radicals, with simazine produced diamino chlorotriazine (Fig. 3d), supporting the possible role of oxygen radicals in the cytochrome P-450 pathway. Adams et al. (1990) also found that phase II products were limited to glutathione conjugates (see Fig. 3).

Summary

Simazine, first introduced in 1956, is a popular agricultural herbicide used to inhibit photosynthesis in broadleaf weeds and grasses. It is a member of the triazine family, and according to its physicochemical properties, it is slightly soluble in water, relatively nonvolatile, capable of partitioning into organic phases, and susceptible to photolysis. Sorption and desorption studies on its behavior in soils indicate that simazine does not appreciably sorb to minerals and has the potential to leach in clay and sandy soils. The presence of organic matter in soils contributes to simazine retention but delays its degradation. The primary sorptive mechanism of simazine to OM has been proposed to be via partitioning and/or by the interaction with functional groups of the sorbent. Farming practices directly influence the movement of simazine in soils as well. Tilled fields lower the runoff of simazine when compared to untilled fields, but tilling can also contribute to its movement into groundwater. Planting cover crops on untilled land can significantly reduce simazine runoff. Such practices are important because simazine and its byproducts have been detected in groundwater in the Netherlands, Denmark, and parts of the U.S. (California, North Carolina, Illinois, and Wisconsin) at significant concentrations. Concentrations have also been detected in surface waters around the U.S. and United Kingdom. Although the physicochemical properties of simazine do not support volatilization, residues have been found in the atmosphere and correlate with its application. Although at low concentrations, simazine has also been detected in precipitation in Pennsylvania (U.S.), Greece, and Paris (France).

Abiotically, simazine can be oxidized to several degradation products. Although hydrolysis does not contribute to the dissipation of simazine, photolysis does. Microbial degradation is the primary means of simazine dissipation, but the process is relatively slow and kinetically controlled. Some bacteria and fungal species capable of utilizing simazine as a sole carbon and nitrogen source at a fast rate under laboratory conditions have been identified. Metabolism of simazine in higher organisms is via cytochrome P-450-mediated oxidation and glutathione conjugation.

Acknowledgments

Support was provided by the Environmental Monitoring Branch of the Department of Pesticide Regulation (CDPR), California Environmental Protection Agency. The statements and conclusions are those of the authors and not necessarily those of CDPR. The mention of commercial products, their source, or their use in connection with materials reported herein is not to be construed as actual or implied endorsement of such products.

References

Adams, NH, Levi, PE, Hodgson, E (1990) In vitro studies of the metabolism of atrazine, simazine, and terbutryn in several vertebrate species. J Agric Food Chem 38:1411–1417.

Ahrens, WH (1994) Herbicide Handbook, 7th Ed. Weed Science Society of America, Champaign, IL.

Albanis, TA, Hela, DG, Sakellarides, TM, Konstantinou, IK (1998) Monitoring of pesticide residues and their metabolites in surface and underground waters of Imathia (N. Greece) by means of solid-phase extraction disks and gas chromatography. J Chromatogr A 823:59–71.

Bacey, J, Starner, K, Spurlock, F (2004) The Occurrence and Concentration of Esfenvalerate and Permethrin in Water and Sediment in the Sacramento and San Joaquin Watersheds. EH 04-01. California Environmental Protection Agency, Sacramento, CA, pp 340–341.

Barriuso, E, Houto, S, Serra-Wittling, C (1997) Influence of compost addition to soil on the behavior of herbicides. Pestic Sci 49:65–75.

Battaglin, WA, Goolsby, DA (1999) Are shifts in herbicide use reflected in concentration changes in Midwestern rivers? Environ Sci Technol 33:2917–2925.

Beltran, J, Gerritse, RG, Hernandez, F (1998) Effect of flow rate on the adsorption and desorption of glyphosate, simazine, and atrazine in columns of sandy soils. Eur J Soil Sci 49:149–156.

Brereton, C, House, WA, Armitage, PD, Wotten, RS (1999) Sorption of pesticides to novel materials: snail pedal mucus and blackfly silk. Environ Pollut 105:55–65.

CDPR (California Department of Pesticide Regulations) (2003a) Summary Report: Complete Report with Summary Data Indexed by Chemical. California Environmental Protection Agency, Sacramento, CA, pp 340–341. http://www.cdpr.ca.gov/docs/pur/pur03rep/chmrpt 03.pdf (verified February 1, 2006).

CDPR (California Department of Pesticide Regulations) (2003b) Top 100 Lists: The Top 100 Pesticides Used Statewide (All Sites Combined) in 2003. California Environmental Protection Agency, Sacramento, CA, p 1. http://www.cdpr.ca.gov/docs/pur/pur03rep/top100_ais.pdf (verified February 1, 2006).

Celis, R, Cornejo, J, Hermosin, MC, Koskinen, WC (1997) Sorption-desorption of atrazine and simazine by model soil colloidal components. Soil Sci Soc Am J 61:436–443.

Celis, R, Cornejo, J, Hermosin, MC, Koskinen, WC (1998) Sorption of atrazine and simazine by model associations of soil colloids. Soil Sci Soc Am J 62:165–171.
Chevreuil, M, Garmouma, M, Teil, MJ, Chesterikoff, A (1996) Occurrence of organochlorines (PCBs, pesticides) and herbicides (triazines, phenylureas) in the atmosphere and in the fallout from urban and rural stations of the Paris area. Sci Total Environ 182:25–37.
Cogger, CG, Bristow, PR, Stark, JD, Getzin, LW, Montgomery, M (1998) Transport and Persistence of pesticide in alluvial soils: I. simazine. J Environ Qual 27:543–550.
Comber, SDW (1999) Abiotic persistence of atrazine and simazine in water. Pestic Sci 55:696–702.
Cook, A, Hutter, R (1984) Deethylsimazine bacterial dechlorination, deamination and complete degradation. J Agric Food Chem 32:581–585.
Cox, L, Hermosín, MC, Cornejo, J (1999) Leaching of simazine in organic-amended soils. Commun Soil Sci Plan 30:1697–1706.
Cox, L, Celis, R, Hermosín, MC, Cornejo, J, Zsolnay, A, Zeller, K (2000a) Effects of organic amendments on herbicide sorption as related to the nature of the dissolved organic matter. Environ Sci Technol 34:4600–4605.
Cox, L, Celis, R, Hermosín, MC, Cornejo, J (2000b) Natural soil colloids to retard simazine and 2,4-D leaching in soil. J Agric Food Chem 48:93–99.
Cox, L, Cecchi, A, Celis, R, Hermosín, MC, Koskinen, WC, Cornejo, J (2001) Effect of exogenous carbon on movement of simazine and 2,4-D in soils. Soil Sci Soc Am J 65:1688–1695.
Cremlyn, RJ (1990) Agrochemicals: Preparation and Mode of Action. J Wiley, West Sussex, UK.
Cruz-Guzmán, M, Celis, R, Hermosín, MC, Cornejo, J (2004) Adsorption of the herbicide simazine by montmorillonite modified with natural organic cations. Environ Sci Technol 38:180–186.
Curren, MSS, King, JW (2001) Solubility of triazines pesticides in pure and modified subcritical water. Anal Chem 73:740–745.
Davis, NM, Lydy, MJ (2002) Evaluating best management practices at an urban golf course. Environ Toxicol Chem 21:1076–1084.
Edwards, CA (1965) Effects of pesticide residues on soil invertebrates and plants. In: 5th Symposium of the British Ecological Society, UK, pp 239–261.
Ertunç, T, Hartlieb, N, Berns, A, Klein, W, Schaeffer, A (2002) Investigation on the binding mechanism of the herbicide simazine to dissolved organic matter in leachates from compost. Chemosphere 49:597–604.
Evgenidou, E, Fytianos, K (2002) Photodegradation of trazine herbicides in aqueous solutions and natural waters. J Agric Food Chem 50:6423–6427.
Filazzola, MT, Sanniono, F, Gianfreda, L (1999) Effect of various pollutants and soil-like constituents on laccase from *Cerrena unicolor*. J Environ Qual 28:1929–1938.
Flury, M (1996) Experimental evidence of transport of pesticides through field soils; a review. J Environ Qual 25:24–45.
García-Valcárcel, AI, Tadeo, JL (1999) Influence of soil moisture on sorption and degradation of hexazinon and simazine in soil. J Agric Food Chem 47:3895–3900.
Ghabbour, SI, Imam, M (1967) The effect of five herbicides on three oligochaete species. Rev Ecol Biol Sol 4:119–122.

Glenn, S, Angle, JS (1987) Atrazine and simazine in runoff from conventional and no-till corn watersheds. Agric Ecosyst Environ 18:273–280.

Glotfelty, DE, Leech, MM, Jersey, J, Taylor, AW (1989) Volatilization and wind erosion of soil surface applied atrazine, simazine, alachlor, and taxaphene. J Agric Food Chem 37:546–551.

Habecker, MA (1989) Environmental Contamination at Wisconsin Pesticide Mixing/Loading Facilities: Case Study, Investigations and Remedial Action Evaluation. Wisconsin Department of Agriculture, Trade and Consumer Protection, Agricultural Resource Management Division, Madison, WI.

Harman-Fetcho, JA, McConnell, LL, Baker, JE (1999) Agricultural pesticides in the Patuxent river, a tributary of the Chesapeake bay. J Environ Qual 28:928–938.

Hartlieb, N, Marschner, B, Klein, W (2001) Transformation of dissolved organic matter (DOM) and ^{14}C-labeled organic contaminants during composting of municipal biowaste. Sci Total Environ 278:1–10.

Hassink, J, Klein, A, Kordel, W, Klein, W (1994) Behavior of herbicides in non-cultivated soils. Chemosphere 28:285–295.

Holden, LR, Graham, JA, Whitmore, RW, Alexander, WJ, Pratt, RW, Liddle, SK, Piper, LL (1992) Results of the national alachlor well water survey. Environ Sci Technol 26:935–943.

Huston, PL, Pignatello, JJ (1999) Degradation of selected pesticide active ingredients and commercial formulations in water by the photo-assisted Fenton reaction. Water Res 33:1238–1246.

Kodama, T, Ding, L, Yoshida, M, Yajima, M (2001) Biodegradation of an *s*-triazine herbicide, simazine. J Mol Catal B Enzym 11:1073–1078.

Kosanke, GJ, Schwippert, WW, Beneke, TW (1988) The impairment of mobility and development in freshwater snails (*Physafontinalis* and *Lymnaea stagnalis*) caused by herbicides. Comp Biochem Phys C 90:373–380.

Laabs, V, Amelung, W, Fent, G, Zech, W, Kubiak, R (2002) Fate of ^{14}C-labeled soybean and corn pesticides in tropical soils of Brazil under laboratory conditions. J Agric Food Chem 50:4619–4627.

Lagas, P, Verdam, B, Loch, JPG (1989) Threat to groundwater quality by pesticides in the Netherlands. In: Sahuquillo J, Andreu J, O'Donnell T (eds) Groundwater Management: Quantity and Quality. International Association of Hydrological Sciences, Wallingford, UK, pp 171–180.

Laird, DA, Barriuso, E, Dowdy, RH, Koskinen, WC (1992) Adsorption of atrazine on smectites. Soil Sci Soc Am J 56:62–67.

Lennartz, B, Louchart, X, Voltz, M, Andrieux, P (1997) Diuron and simazine losses to runoff water in Mediterranean vineyards. J Environ Qual 26:1493–1502.

Liu, F, O'Connell, NV (2002) Off-site movement of surface-applied simazine from a citrus orchard as affected by irrigation incorporation. Weed Sci 50:672–676.

Liu, F, O'Connell, NV (2003) Movement of simazine in runoff water and weed control from citrus orchard as affected by reduced rate of herbicide application. Bioresour Technol 86:253–258.

Long, T (1989) Groundwater contamination in the vicinity of agrichemical mixing and loading facilities. In: Illinois Agricultural Pesticides Conference: Proceedings. University of Illinois College of Agricultural Consumer and Environmental Services, Urbana-Champaign, IL, pp 139–149.

Louchart, X, Voltz, M, Andrieux, P, Moussa, R (2001) Herbicide transport to surface waters at field and watershed scales in a Mediterranean vineyard area. J Environ Qual 30:982–991.
Martin-Montalvo, D, Mengs, G, Ferrer, E, Allende, JL, Alonso, R, Martin, M (1997) Simazine degradation by immobilized and suspended soil bacterium. Int Biodeterior Biodegrad 40:93–99.
Matsui, Y, Knappe, DRU, Iwaki, K, Ohira, H (2002) Pesticide adsorption by granular activated carbon adsorbers. 2. Effects of pesticide and natural organic matter characteristics on pesticide breakthrough curves. Environ Sci Technol 36:3432–3438.
Melnikov, NN (1971) Chemistry of pesticides. In: Gunther FA, Gunther JD (eds) Residue Reviews. Springer-Verlag, New York, 36:435.
Montgomery, JH (1997) Agrochemicals Desk Reference. CRC Press, Boca Raton, FL, pp 162–168.
Mougin, C, Laugero, C, Asther, M, Dubroca, J, Frasse, P (1994) Biotransformation of the herbicide atrazine by the white rot fungus *Phanerochete chrysosporium*. Appl Environ Microbiol 60:705–708.
Mougin, C, Laugero, C, Asther, M, Chaplain, V (1997) Biotransformation of *s*-traizine herbicides and related degradation products in liquid cultures by the white rot fungus *Phanerochaete chrysosporium*. Pestic Sci 49:169–177.
Noblet, JA, Smith, LA, Suffet, IHM (1996) Influence of natural dissolved organic matter, temperature, and mixing on the aboitic hydrolysis of triazines and organophosphate pesticides. J Agric Food Chem 44:3685–3693.
Oliveira, RS Jr, Koskinen, WC, Ferriera, FA (2001) Sorption and leaching potential of herbicides on Brazilian soils. Weed Res 41:97–110.
Power, M, Attrill, MJ, Thomas, RM (1999) Trends in agricultural pesticide (atrazine, lindane, simazine) concentrations in the Thames estuary. Environ Pollut 104:31–39.
Reddy, KN, Singh, M, Alvia, AK (1992) Sorption and leaching of bromacil and simazine in Florida flatwoods soils. Bull Environ Contam Toxicol 48:662–670.
Ritter, WF, Chirnside, AEM, Scarborough, RW (1996) Movement and degradation of triazines, alachlor, and metolachlor in sandy soils. J Environ Sci Health A 31:2699–2721.
Rivas, J, Beltrán, FJ, García-Araya, JF, Navarrete, V (2001) Simazine removal from water in a continuous bubble column by O_3 and O_3/H_2O_2. J Environ Sci Health B 36:809–819.
Sanders, HO (1969) Toxicities of pesticides to the crustacean *Gammarus lacustris*. Tech Pap Bur Sport Fish Wildl 25:1–17.
Sannino, F, Filazzola, MT, Violante, A, Gianfreda, L (1999a) Adsorption-desorption of simazine on montmorillonite coated by hydroxyl aluminum species. Environ Sci Technol 33:4221–4225.
Sannino, F, Filazzola, MT, Violante, A, Gianfreda, L (1999b) Fate of herbicides influenced by biotic and abiotic interactions. Chemosphere 39:333–341.
Sawunyama, P, Bailey, GW (2002) Computational chemistry study of the environmentally important acid-catalyzed hydrolysis of atrazine and related 2-chloro-*s*-triazines. Pestic Manage Sci 58:759–768.
Shertzer, RH, Hall, DW, Steffy, SA, Kime, RA (1998) Relationships between land uses and rainwater quality in a south central Pennsylvania watershed. J Am Water Res Assoc 34:13–26.

Spliid, NH, Koppen, B (1998) Occurrence of pesticides in Danish shallow ground water. Chemosphere 37:1307–1316.

Spurlock, F, Burow, K, Dubrovsky, N (2000) Chlorofluorocarbon dating of herbicide-containing well waters in Fresno and Tulare counties, California. J Environ Qual 29:474–483.

Stearman, GK, Wells, JM, Cripps, RW (1997) Leaching and runoff of simazine, 2–4-D, and bromide from nursery plots. J Soil Water Conserv 52:137–144.

Strong, LC, Rosendahl, C, Johnson, G, Sadwosky, MJ, Wackett, LP (2002) *Arthrobacter aurescens* TC1 metabolizes diverse *s*-triazine ring compounds. Appl Environ Microbiol 68:5973–5980.

Tennant, AH, Peng, BC, Kligerman, AD (2001) Genotoxicity studies of three triazine herbicides: in vivo studies using the alkaline single cell gel (SCG) assay. Mutat Res 493:1–10.

Troiano, J, Garretson, C (1998) Movement of simazine in runoff water from citrus orchard row middles as affected by mechanical incorporation. J Environ Qual 27:488–494.

Troiano, J, Nordmark, C (2002) Memo: distribution of triazine residues in wells in relation to current and proposed maximum contaminant levels (MCLs). Department of Pesticide Regulation, Cal/EPA, Sacramento, CA. http://www.cdpr.ca.gov/docs/empm/pubs/memos.htm (verified February 1, 2006).

Troiano, J, Weaver, D, Marade, J, Spurlock, F, Pepple, M, Nordmark, C, Bartkowiak, D (2001) Summary of well water sampling in California to detect pesticide residues resulting from nonpoint-source applications. J Environ Qual 30:448–459.

U.S. EPA (2003) Consumer factsheet on simazine. Available at: http://www.epa.gov/safewater/contaminants/dw_contamfs/simazine.html (verified September 25, 2006)

Vencill, WK, Armbrust, K, Hancock, GH, Johnson, D, McDonald, G, Kintner, D, Lichtner, F, McLean, H, Reynolds, J, Rushing, D, Senseman, S, Wauchope, D (2002) Simazine. In: Vencill WK (ed) Herbicide Handbook, 8th Ed. Weed Science Society of America, Lawrence, KS, pp 397–399.

Verschueren, K (1984) Handbook of Environmental Data on Organic Chemicals, 2nd Ed. Van Nostrand Reinhold, New York, pp 363–364.

Vink, JPM, van der Zee, SEATM (1997a) Effect of oxygen status on pesticide transformation and sorption in undistributed soil and lake sediment. Environ Toxicol Chem 16:608–616.

Vink, JPM, van der Zee, SEATM (1997b) Pesticide biotransformation in surface waters: multivariate analyses of environmental factors at field sites. Water Res 31:2858–2868.

Wade, HF, York, AC, Morey, E, Padmore, JM, Rudo, KM (1998) The impact of pesticide use on groundwater in North Carolina. J Environ Qual 27: 1018–1026.

Wang, Y-S, Duh, J-R, Lin, K-Y, Chen, Y-L (1996) Movement of three *s*-triazine herbicides atrazine, simazine, and ametryn in subtropical soils. Bull Environ Contam Toxicol 57:743–750.

Ware, GW, Whitacre, DM (2004) The Pesticide Book, 6th Ed. MeisterPRO Publications, Willoughby, OH.

Wauchope, RD (1978) The pesticide content of surface waters draining from agricultural fields: a review. J Environ Qual 7:459–472.

Wauchope, RD, Buttler, TM, Hornsby, AG, Augustijn-Beckers, PWM, Burt, JP (1992) The SCS/ARS/CES pesticide properties database for environmental decision-making. Rev Environ Contam Toxicol 123:1–164.
Wilson, PC, Whitwell, T, Klaine, SJ (1999) Phytotoxicity, uptake, and distribution of [^{14}C] simazine in *canna hybrida* "Yellow King Humbert". Environ Toxicol Chem 18:1462–1468.

Manuscript received February 4; accepted February 7, 2006.

Rev Environ Contam Toxicol 189:25–41

Ethanol Production: Energy, Economic, and Environmental Losses

David Pimentel, Tad Patzek, and Gerald Cecil

Contents

I. Introduction 25
II. Corn Use in Ethanol Production 26
A. Energy Inputs in Corn Production 26
B. Energy Inputs in Fermentation/Distillation 28
C. Net Energy Yield 29
D. Economic Costs 29
E. Cornland Use 30
F. By-Products 30
G. Sugarcane Use in Ethanol Production 31
III. Ecological Issues 31
A. Cropland Use 31
B. Environmental Impacts 32
C. Negative or Positive Energy Return? 33
D. Food Security 34
E. Food versus Fuel Issue 35
Summary 35
Acknowledgments 36
References 37

I. Introduction

The supply of "conventional" oil is projected to peak before 2010, and its decline thereafter cannot be compensated fully by other liquid fuels (Youngquist and Duncan 2003). The United States critically needs to develop liquid fuel replacements for oil in the near future. The search for alternative liquid fuels has focused on the use of biomass.

Communicated by David M. Whitacre.

D. Pimentel (✉)
College of Agriculture and Life Sciences, 5126 Comstock Hall, Cornell University, Ithaca, NY 14853, U.S.A.

T. Patzek
Department of Civil and Environmental Engineering, 425 Davis Hall, University of California, Berkeley, CA 94720, U.S.A.

G. Cecil
Department of Physics and Astronomy, 290 Phillips Hall, University of North Carolina, Chapel Hill, NC 27599, U.S.A.

Biomass is green plant material, such as corn, soybeans, sugarcane, and trees. All kinds of biomass convert solar energy into plant material, but this conversion requires suitable soil, nutrients, and freshwater. Then, in the conversion of the biomass into liquid fuel, water, microorganisms, and more energy are required. Andrew Ferguson (2004) makes an astute observation that the proportion of the sun's energy that is converted into useful ethanol, even using very positive energy data, only amounts to 1 part per 1,000, or 0.1% of the solar energy.

Two early studies by the U.S. Department of Energy (USDOE) concerning ethanol production using corn for liquid fuels from biomass reported a negative energy return (ERAB 1980, 1981). These reports were reviewed by 26 expert U.S. scientists independent of the USDOE; their findings concluded that the conversion of corn into ethanol energy was negative, and these findings were unanimously approved. Since then, other investigations have confirmed these earlier findings (Pimentel and Patzek 2005).

In this analysis, the most recent scientific data for corn production and for fermentation/distillation were used. All current fossil energy inputs used in corn production and for the fermentation/distillation were included to determine the entire energy cost of ethanol production. Additional costs to consumers include federal and state subsidies, plus costs associated with environmental pollution and/or degradation that occur during the entire production process. Economic and human food supply issues are discussed. In addition, studies that contrast with the conclusion of this study are evaluated.

II. Corn Use in Ethanol Production

A. Energy Inputs in Corn Production

The conversion of corn and other food/feed crops into ethanol by fermentation is a well-known and established technology. The ethanol yield from a large production plant is about 1L ethanol from 2.69kg corn grain (Pimentel and Patzek 2005).

The production of corn in the U.S. requires a significant energy and dollar investment for the 14 inputs, including labor, farm machinery, fertilizers, irrigation, pesticides, and electricity (Table 1). To produce an average corn yield of 8,781kg/ha [140 bushels/acre (bu/ac)] using up-to-date production technologies requires the expenditure of about 7.5 million kcal energy input, mostly oil and natural gas, as listed in Table 1; this is the equivalent of about 854L oil equivalents expended per hectare of corn. The production costs total about $892/ha for the 8,781kg/ha or approximately $0.10/kg ($2.58/ bu) of corn produced.

Full irrigation, when there is insufficient or no rainfall, requires about 100cm water per growing season. Because only about 15% of U.S. corn

Table 1. Energy inputs and costs of corn production per hectare in the United States.

Inputs	Quantity	kcal ×1,000	Costs ($)
Labor	11.4 hr[a]	462[b]	148.20[c]
Machinery	18 kg[d]	333[e]	68.00[f]
Diesel	88 L[g]	1,003[h]	34.76
Gasoline	40 L[i]	405[j]	20.80
Nitrogen	155 kg[k]	2,480[l]	85.25[m]
Phosphorus	79 kg[n]	328[o]	48.98[p]
Potassium	84 kg[q]	274[r]	26.04[s]
Lime	1,120 kg[t]	315[u]	19.80
Seeds	21 kg[v]	520[w]	74.81[x]
Irrigation	8.1 cm[y]	320[z]	123.00[aa]
Herbicides	6.2 kg[bb]	620[cc]	124.00
Insecticides	2.8 kg[dd]	280[cc]	56.00
Electricity	13.2 kWh[ee]	34[ff]	0.92
Transport	204 kg[gg]	169[hh]	61.20
Total		7,543	891.76
Corn yield 8,781 kg/ha[ii]		31,612 (kcal input : output, 1 : 4.19)	

[a]NASS (2003).
[b]It is assumed that a person works 2000 hr/yr and utilizes an average of 8000 L oil equivalents/yr.
[c]It is assumed that labor is paid $13/hr.
[d]Energy costs for farm machinery that was obtained from agricultural engineers: tractors, harvesters, plows, and other equipment that last about 10 years and are used on 160 ha/yr. These data were prorated per year per hectare (Pimentel and Patzek 2005).
[e]Prorated per hectare and 10-yr life of the machinery. Tractors weigh about 10 t and harvesters about 10 t (International Harvester 2006), plus plows, sprayers, and other equipment.
[f]Hoffman et al. (1994).
[g]Wilcke and Chaplin (2000).
[h]Input 11,400 kcal/L.
[i]Estimated.
[j]Input 10,125 kcal/L.
[k]NASS (2003).
[l]Patzek (2004).
[m]Cost $0.55/kg.
[n]NASS (2003).
[o]Input 4,154 kcal/kg.
[p]Cost $0.62/kg.
[q]NASS (2003).
[r]Input 3,260 kcal/kg.
[s]Cost $0.31/kg.
[t]Brees (2004).
[u]Input 281 kcal/kg.
[v]Pimentel and Pimentel (1996).
[w]Pimentel and Pimentel (1996).
[x]USDA (1997b).
[y]USDA (1997a).
[z]Batty and Keller (1980).
[aa]Irrigation for 100 cm water/ha costs $1,000 (Larsen et al. 2002).
[bb]Larson and Cardwell (1999).
[cc]Input 100,000 kcal/kg of herbicide and insecticide.
[dd]USDA (2002).
[ee]USDA (1991).
[ff]Input 860 kcal/kWhr; requires 3 kWhr thermal energy to produce 1 kWhr electricity.
[gg]Goods transported include machinery, fuels, and seeds that were shipped an estimated 1,000 km.
[hh]Input 0.83 kcal/kg/km transported.
[ii]USCB (2004–2005). There is a need to look at average crop yield over 3–5 years, not record peak years, as a base for fuel policy decisions.

production currently is irrigated (USDA 1997a), only 8.1 cm/ha of irrigation was included for the growing season. On average, irrigation water is pumped from a depth of 100 m (USDA 1997a). On this basis, the average energy input associated with irrigation is 320,000 kcal/ha (see Table 1).

B. Energy Inputs in Fermentation/Distillation

The average costs in terms of energy and dollars for a large (245–285 million L/yr) modern ethanol plant are listed in Table 2. In the fermentation/distillation process, the corn is finely ground and approximately 15 L water is added per 2.69 kg ground corn. After fermentation, to obtain 1 L 95% pure ethanol from the mixture of 8% ethanol and 92% water, 1 L ethanol must be extracted from the approximately 13 L of the ethanol/water mixture.

Table 2. Inputs per 1,000 L 99.5% ethanol produced from corn.[a]

Inputs	Quantity	kcal ×1,000	Cost ($)
Corn grain	2,690 kg[b]	2,314[b]	273.62
Corn transport	2,690 kg[b]	322[c]	21.40[d]
Water	15,000 L[e]	90[f]	21.16[g]
Stainless steel	3 kg[i]	165[p]	10.60[d]
Steel	4 kg[i]	92[p]	10.60[d]
Cement	8 kg[i]	384[p]	10.60[d]
Steam	2,546,000 kcal[j]	2,546[j]	21.16[k]
Electricity	392 kWhr[j]	1,011[j]	27.44[l]
95% ethanol to 99.5%	9 kcal/L[m]	9[m]	0.60
Sewage effluent	20 kg BOD[n]	69[h]	6.00
Distribution	331 kcal/L[q]	331	20.00[q]
Total		7,333	423.18

BOD, biological oxygen demand.
[a]Output: 1 L of ethanol = 5,130 kcal.
[b]Data from Table 1.
[c]Calculated for 144 km round trip.
[d]Pimentel (2003).
[e]15 L water mixed with each kilogram grain.
[f]Pimentel et al. (2004b).
[g]Pimentel et al. (2004b).
[h]4 kWhr energy required to process 1 kg BOD (Blais et al. 1995).
[i]Estimated.
[j]Illinois Corn (2004).
[k]Calculated based on the price of natural gas.
[l]$0.07/kWhr (USCB 2004–2005).
[m]95% ethanol converted to 99.5% ethanol for addition to gasoline (T. Patzek, personal communication, University of California, Berkeley, 2004).
[n]20 kg BOD/1,000 L ethanol produced (Kuby et al. 1984).
[p]Newton (2001).
[q]DOE (2002).

Although ethanol boils at about 78°C and water boils at 100°C, the ethanol is not extracted from the water in just one distillation, which obtains 95% pure ethanol (Maiorella 1985; Wereko-Brobby and Hagan 1996; S. Lamberson, personal communication, Cornell University, 2000). To be mixed with gasoline, the 95% ethanol must be further processed and more water removed, requiring additional fossil energy input to achieve 99.5% pure ethanol (Table 2). Thus, a total of about 12 L wastewater must be removed per 1 L ethanol produced, and this relatively large amount of sewage effluent must be disposed of at an energy, economic, and environmental cost.

To produce a liter of 99.5% ethanol uses 43% more fossil energy than the energy produced as ethanol and costs \$0.42/L(\$1.59/gal) (see Table 2). The corn feedstock requires more than 33% of the total energy input. In Table 2, which presents distillation/fermentation processes, the corn grain and transport inputs account for 36% of the total energy spent, and the remaining energy input is 64%. In this analysis, the total cost, including the energy inputs for the fermentation/distillation process and the apportioned energy costs of the stainless steel tanks and other industrial materials, is \$423.18/1,000 L ethanol produced (Table 2).

C. Net Energy Yield

The largest energy inputs in corn-ethanol production are for producing the corn feedstock, plus the steam energy and electricity used in the fermentation/ distillation process. The total energy input to produce 1 L ethanol is 7,333 kcal (Table 2). However, 1 L ethanol has an energy value of only 5,130 kcal. Based on a net energy loss of 2,203 kcal ethanol produced, 43% more fossil energy is expended than is produced as ethanol. The energy deficit might be reduced if ethanol producers were able to provide the low-pressure steam required for distillation from cogeneration power facilities or from solar thermal inputs.

D. Economic Costs

Current ethanol production technology uses more fossil fuel and costs substantially more to produce in dollars than its energy value is worth on the market. Clearly, without the more than \$3 billion federal and state government yearly subsidies, U.S. ethanol production would be reduced or cease, confirming the basic fact that ethanol production is uneconomical (National Center for Policy Analysis 2002).

Federal and state subsidies for ethanol production that total more than \$7 per bushel of corn are mainly paid to large corporations (McCain 2003), while corn farmers are receiving a maximum of only an added \$0.02 per bushel for their corn (\$6.90/ha or \$2.80/A) in the subsidized corn ethanol production system (Pimentel and Patzek 2005). Senator McCain reports that direct subsidies for ethanol, plus the subsidies for corn grain, amount to \$0.79/L (McCain 2003).

If the production costs of a liter of ethanol were added to the tax subsidy cost, then the total cost per liter of ethanol would be \$1.21. Because of the relatively low energy content of ethanol, 1.6L ethanol has the energy equivalent of 1L gasoline. Thus, the cost of producing an equivalent amount of ethanol equal to 1L gasoline is \$1.94 (\$7.32/gal gasoline); this is more than \$0.53/L, the current cost of producing 1L gasoline.

Unfortunately, the costs to the American consumer are greater than the \$8.4 billion/yr expended to subsidize corn ethanol because diverting the required corn feedstock from livestock increases corn prices for livestock producers. The National Center for Policy Analysis (2002) estimate is that ethanol production is adding more than \$1 billion/yr to the cost of beef production for consumers. Given that about 70% of the current corn grain harvest is fed to U.S. livestock (USDA 2003), doubling or tripling ethanol production can be expected to increase corn prices further for beef production and for other livestock products, including milk and eggs, and ultimately to increase costs to the consumer. Therefore, in addition to paying the \$8.4 billion in taxes for ethanol and corn subsidies, consumers are expected to face significantly higher meat, milk, and egg prices in the marketplace.

E. Cornland Use

Currently, about 17.0 billion L ethanol (4.5 billion gal) is produced in the U.S. each year (DOE 2005). The total liquid fuels used in the U.S. were about 1,200 billion L in 2003 (Pimentel et al. 2004a; USCB 2004–2005). Therefore, 17.0 billion L ethanol (energy equivalent to 11.2 billion L vehicle liquid fuel) provides only 1% of the liquid fuel utilized by the U.S. each year. To produce this 17.0 billion L ethanol, about 5.1 million ha or 18% of U.S. corn is used. Expanding corn-ethanol production to 100% of U.S. corn production would provide just 6% of the liquid fuel of the U.S.

F. By-Products

The energy and dollar costs of producing ethanol can be offset partially by by-products, such as the dry distillers grains (DDG) made from dry-milling of corn. From about 10kg corn feedstock, about 3.3kg DDG with 27% protein content can be harvested (Stanton 1999). This DDG is suitable for feeding cattle, which are ruminants, but has only limited value for feeding hogs and chickens. In practice, this DDG is generally used as a substitute for soybean feed, which contains 49% protein (Stanton 1999). However, soybean production for livestock feed is more energy efficient than corn production because little or no nitrogen fertilizer is needed for the production of this legume (Pimentel et al. 2002). In practice, only 2.1 kg soybean protein provides the equivalent nutrient value of 3.3kg DDG. Thus, the credit fossil energy/L ethanol produced is about 445kcal (Pimentel et al. 2002). Factoring this credit for a nonfuel source into the production of ethanol reduces the negative energy balance for ethanol production from 43% to 28% (see Table 2). The high energy credits for DDG given by some are unrealistic because the production of livestock feed from ethanol is uneconomical given the high costs of fossil energy plus the costs of soil depletion to the farmer (Patzek 2004). The resulting overall energy output–input comparison remains negative even with the credits for the DDG by-product.

G. Sugarcane Use in Ethanol Production

Proponents of ethanol point to the production of ethanol in Brazil, which currently is the largest producer of ethanol in the world. Brazil uses sugarcane to produce ethanol, and sugarcane is a more efficient feedstock for ethanol production than corn grain (Pimentel and Pimentel 1996). However, because the energy balance is negative, the Brazilians subsidize their ethanol industry. Initially the government was selling ethanol to the public for \$0.22/L, but it cost the government \$0.33/L to produce (Pimentel 2003). Because of other economic priorities in Brazil, the government has abandoned directly subsidizing ethanol (Spirits Low 1999; Coelho et al. 2001). The ethanol industry is still being subsidized, but the consumer is paying this subsidy directly at the pump (Pimentel 2003). The subsidy is estimated to be about 50% for ethanol production (CIA 2005). In addition, there are serious ecological concerns related to ethanol production in Brazil, which include the removal of mature forests for more sugarcane plantations and increased soil erosion associated with the culture of sugarcane (Azar et al. 2006).

III. Ecological Issues

A. Cropland Use

When considering the advisability of producing sufficient ethanol for automobiles, the availability of cropland required to grow sufficient corn to fuel each automobile is critical. For the sake of argument we use Shapouri's (Shapouri et al. 2002, 2004) optimistic suggestion that all natural gas and electricity inputs be ignored and only gasoline and diesel fuel inputs be assessed. Based on Shapouri's input–output data, 2,929 L ethanol is produced/corn ha. When equated to gasoline, this ethanol has the same energy as 1,890 L gasoline. An average U.S. automobile travels more than 10,000 miles/yr and uses about 1,890 L gasoline/yr (USCB 2004–2005). To replace this gasoline usage with ethanol, about 1 ha corn would have to be grown. Consider that at present 0.5 ha U.S. cropland is used to feed each person a diverse and nutritious diet (USCB 2004–2005). Therefore, even using Shapouri's optimistic energy accounting data, to fuel one automobile with ethanol, as a substitute for the yearly use of gasoline for 1 yr, two times more cropland would be required for corn production and ethanol production than now is required to feed one person!

Worldwide, for ethanol to replace gasoline, about 2.4 billion ha cropland planted to corn would be required, which represents 60% more cropland than exists in the world (A.R.B. Ferguson, personal communication, Optimum Population Trust, November 6, 2005).

B. Environmental Impacts

Some of the economic and energy contributions of the by-products are negated by the widespread environmental pollution problems associated with ethanol production. First, corn production causes more soil erosion than any other U.S. crop (Pimentel et al. 1995; NAS 2003). In addition, corn production uses more herbicides and insecticides and nitrogen fertilizer than any other crop produced in the U.S., and these chemicals may invade groundwater and surface water, thereby causing more water pollution than any other crop (NAS 2003).

As mentioned, the production of 1 L ethanol requires 1,700 L freshwater, both for corn production and for the fermentation/distillation processing of ethanol (Pimentel et al. 2004b). In some Western irrigated corn acreage, e.g., some regions of Arizona, groundwater is being pumped 10 times faster than the natural recharge of the aquifers (Pimentel et al. 2004b).

All these factors confirm that the environmental and agricultural system in which U.S. corn is being produced is experiencing major degradation. Further, it substantiates the conclusion that the U.S. corn production system, and indeed the entire ethanol production system, is not environmentally sustainable now or for the future, unless major changes are made in the cultivation of this major food/feed crop. Because corn is raw material for ethanol production, it cannot be considered a renewable energy source.

Furthermore, pollution problems associated with the production of ethanol at the chemical plant sites are emerging. The EPA (2002) already has issued warnings to ethanol plants to reduce their air pollution emissions or be shut down. Another pollution problem concerns the large amounts of wastewater produced by each ethanol plant. As noted, for each liter of ethanol produced using corn, about 12 L wastewater is produced. This polluting wastewater has a biological oxygen demand (BOD) of 18,000–37,000 mg/L depending of the type of plant (Kuby et al. 1984). The cost of processing this sewage in terms of energy (4 kWhr/kg BOD) was included in the cost of producing ethanol (see Table 2).

Reports confirm that ethanol use contributes to air pollution problems when burned in automobiles (Youngquist 1997; Hodge 2002, 2003, 2005; Niven 2005). The use of both fossil fuels and ethanol releases significant quantities of pollutants to the atmosphere. Furthermore, carbon dioxide emissions released from burning these fossil fuels contribute to global warming and are a serious concern (Schneider et al. 2002). When all the air pollutants associated with the entire ethanol production system are considered, the evidence confirms that ethanol production contributes to the already serious U.S. air pollution problem (Youngquist 1997; Pimentel and Patzek 2005). Investments to control these air pollution problems in the ethanol production plant are possible but will add to the significant production costs of ethanol.

C. Negative or Positive Energy Return?

Shapouri (Shapouri et al. 2004) of the USDA is now reporting a net energy positive return of 67%, whereas in this chapter, we report a negative 43% deficit. In their earlier report, Shapouri et al. (2002) reported a net energy positive return of 34%. Why did ethanol production net return for the USDA nearly double in 2 yr, while corn yields in the U.S. declined 6% during that period (USDA 2002, 2003)? The Shapouri results need to be examined and explained.

1. Shapouri et al. (2004) omit several inputs. For instance, all the energy required to produce and repair farm machinery and the fermentation-distillation equipment is not included. All corn production in the U.S. uses an abundance of farm machinery, including tractors, planters, sprayers, harvesters, and other equipment. These uses contribute substantial energy inputs in corn ethanol production, even when allocated on a life-cycle basis.
2. Shapouri et al. used corn data from only 9 states, compared to this analysis which includes corn data from 50 states most of which have ethanol plants.
3. Shapouri et al. reported a net energy return of 67% after including the co-products, primarily dried distillers grain (DDG) used to feed cattle. These co-products are not fuel! Giampietro et al. (1997) observed that although the by-product DDG may be considered as a positive output in the calculation of the output/input energy ratio in ethanol production, in a large-scale production of ethanol fuel, the DDG would be many times the commercial livestock feed needs each year in the U.S. (Giampietro et al. 1997). It follows then that in a large-scale biofuel production, the DDG could become a serious waste disposal problem and increase the energy costs.

Farrell et al. (2006) report a small positive energy return for corn ethanol but less than half that suggested by Shapouri et al. (2004). The Farrell et al. paper includes the following questionable assumptions:

1. By-products are not the same as "whole corn" for livestock feed.
2. An excessive energy value is allocated for the by-products that would be used to displace the cheaper, more nutritious soybean meal for livestock.

3. Labor of the farmer on the farm was not included.

4. Energy costs for farm machinery were greatly reduced without documentation. These inputs are substantial and were prorated per year per hectare (Pimentel and Patzek 2005).

5. Conservation tillage does save tractor fuel. However, the practice requires the use of significantly more hybrid corn seed, nitrogen fertilizer, herbicides, insecticides, and sometimes rodenticides and molluscicides. All these items require fossil energy for production and application.

6. The only environmental factor mentioned was global warming. Not considered were soil erosion, water use, insecticides, herbicides, and nitrogen fertilizer, which are serious environmental pollutants.

7. We note that Farrell in World Environment News (2006) is quoted saying that it is "possible to put ethanol in a car and run it, but making ethanol using current technology is expensive and contributes to pollution and greenhouse gases." This conclusion is opposite from that of Farrell et al. (2006).

8. In a press release (Jan. 16, 2006, UC Berkeley) one of the authors (Kammen) stated that ethanol could replace 20%–30% of fuel use in the U.S. The 17.0 billion L ethanol currently being produced is using 18% of all U.S. corn production; but this represents less than 1% of U.S. oil use. If 100% of U.S. corn were converted to ethanol it would provide only 6% of current U.S. vehicle fuel use.

In contrast to the USDA and Farrell et al. studies, numerous scientific studies have concluded that corn ethanol production does not provide a net energy balance, that ethanol is not a renewable energy source, and is not an economical fuel; furthermore, its production and use contribute to air, water, and soil pollution and to global warming (Ho 1989; Giampietro et al. 1997; Youngquist 1997; Pimentel 1998, 2001, 2003; NPRA 2002; Croysdale 2001; CalGasoline 2002; Lieberman 2002; Hodge 2002, 2003, 2005; Ferguson 2003, 2004; Patzek 2004; Pimentel and Patzek 2005; Brown 2005; Anthrop 2005; Transportation Research Board 2006; Hassett 2006).

Basically the major problem with corn and all other biomass crops is that they collect on average only 0.1%–0.2% of the solar energy per year. At a fairly typical gross yield of 3,000 L ethanol/ha/yr, the power density achieved is only 2.1 kW/ha, as compared with the gross power density achieved via oil, after delivery for use, which is of the order of 2,000 kw/ha. (A.R.B. Ferguson, personal communication, Optimum Population Trust, November 6, 2005). If all current 28 million ha of corn production were to be devoted only to growing corn for ethanol, this acreage would supply only about 6% of U.S. liquid fuel needs (USDA 2003).

D. Food Security

At present, world agricultural land supplies more than 99% of all world food calories, whereas aquatic ecosystems supply less than 1% (FAO 2001).

Worldwide, during the last decade, per capita available cropland decreased 20% and irrigation land 12% (Brown 1997). Furthermore, per capita grain production has been decreasing, in part due to increases in the world population (Worldwatch Institute 2001). Worldwide, diverse cereal grains, including corn, make up 80% of the food of the human food supply (Pimentel and Pimentel 1996).

The current food shortages throughout the world call attention to the importance of continuing U.S. exports of corn and other grains for human food. During the past 10 yr, U.S. corn and other grain exports have nearly tripled, increasing U.S. export trade by about $3 billion/yr year (USCB 2004–2005).

The expanding world population, which now numbers 6.5 billion, further complicates and stresses the food security problem now and for the future (PRB 2005). Almost a quarter million people are added each day to the world population, and each of these human beings requires adequate food. Currently, malnourished people in the world number about 3.7 billion (WHO 2000), the largest number and proportion ever reported in history. Malnourished people are highly susceptible to various serious diseases, and this concern is reflected in the rapid rise in the number of seriously infected people in the world, with diseases such as tuberculosis, malaria, and Auto-Immune Deficiency Syndrome (AIDS), as reported by the World Health Organization (Kim 2002; Pimentel et al. 2006).

E. Food Versus Fuel Issue

Using corn, a basic human food resource, for ethanol production raises ethical and moral issues (Wald 2006). Expanding ethanol production entails diverting valuable cropland from the production of corn needed to nourish people. The energetic and environmental aspects, as well as the moral and ethical issues, also deserve serious consideration. With oil and natural gas shortages now facing the U.S., ethanol production is forcing the U.S. to import more oil and natural gas to produce ethanol and other biofuels (Pimentel and Patzek 2005).

Furthermore, increasing oil and natural gas imports drives up the price of oil and gas; this is especially critical for the poor in developing countries of the world. The impact is documented by the fact that worldwide per capita fertilizer use has been declining for the last decade (Worldwatch Institute 2001).

Summary

The prime focus of ethanol production from corn is to replace the imported oil used in American vehicles, without expending more fossil energy in ethanol production than is produced as ethanol energy.

In a thorough and up-to-date evaluation of all the fossil energy costs of ethanol production from corn, every step in the production and conversion process must be included. In this study, 14 energy inputs in average U.S. corn production are included. Then, in the fermentation/distillation operation, 9 more identified fossil fuel inputs are included. Some energy and economic credits are given for the by-products, including dried distillers grains (DDG).

Based on all the fossil energy inputs, a total of 1.43 kcal fossil energy is expended to produced 1 kcal ethanol. When the energy value of the DDG, based on the feed value of the DDG as compared to that of soybean meal, is considered, the energy cost of ethanol production is reduced slightly, to 1.28 kcal fossil energy input per 1 kcal ethanol produced.

Several proethanol investigators have overlooked various energy inputs in U.S. corn production, including farm machinery, processing machinery, and the use of hybrid corn. In other studies, unrealistic, low energy costs were attributed to such inputs as nitrogen fertilizer, insecticides, and herbicides. Controversy continues concerning the energy and economic credits that should be assigned to the by-products.

The U.S. Department of Energy reports that 17.0 billion L ethanol was produced in 2005. This represents only less than 1% of total oil use in the U.S. These yields are based on using about 18% of total U.S. corn production and 18% of cornland. Because the production of ethanol requires large inputs of both oil and natural gas in production, the U.S. is importing both oil and natural gas to produce ethanol.

Furthermore, the U.S. Government is spending about $3 billion annually to subsidize ethanol production, a subsidy of $0.79/L ethanol produced. With the subsidy, plus the cost of production, the cost of ethanol is calculated to be $1.21/L. The subsidy for a liter of ethanol is 45-times greater than the subsidy per liter of gasoline.

The environmental costs associated with producing ethanol are significant but have been ignored by most investigators in terms of energy and economics. The negative environmental impacts on cropland, and freshwater, as well as air pollution and public health, have yet to be carefully assessed. These environmental costs in terms of energy and economics should be calculated and included in future ethanol analyses.

General concern has been expressed about taking 18% of U.S. corn, and more in the future, to produce ethanol for burning in automobiles instead of using the corn as food for the many malnourished people in the world. The World Health Organization reports that more than 3.7 billion humans are currently malnourished in the world – the largest number ever in history.

Acknowledgments

We wish to thank the following people for their helpful comments and suggestions on earlier drafts of the manuscript: A. Ferguson, Energy

Specialist, Optimum Population Trust, Manchester, UK; D. Cooke, Department of Chemistry, Cornell University, Ithaca, NY; P. Harriot, Department of Chemical Engineering, Cornell University, Ithaca, NY; M. Giampietro, International Nutrition Institute, Rome, Italy; S. Ulgiati, Department of Chemistry, University of Siena, Italy; P. Weisz, Department of Chemical Engineering, Pennsylvania State University, State College, PA; E. Kessler, Department of Meteorology, University of Oklahoma, Norman, OK; M. Pimentel, Division of Nutritional Sciences, Cornell University, Ithaca, NY; A. Wilson, Research Assistant, Cornell University, Ithaca, NY; and David Hammer, Professor of Computer Science and Engineering, Cornell University, Ithaca, NY.

References

Anthrop, DF (2005) Biomass potential. Oil Gas J September 5:2.

Azar, C, Berndes, G, Hansson, J, Grahn, M (2006) On Brazilian ethanol and the ecological footprint. BioScience 56(1):5.

Batty, JC, Keller, J (1980) Energy requirements for irrigation. In: Pimentel D (ed) Handbook of Energy Utilization in Agriculture. CRC Press, Boca Raton, FL, pp. 35–44.

Blais, JF, Mamouny, K, Nlombi, K, Sasseville, JL, Letourneau, M (1995) Les mesures deficacite energetique dans le secteur de leau. In: Sassville JL, Balis JF (eds) Les Mesures deficacite Energetique pour Lepuration des eaux Usees Municipales. Scientific Report 405, vol 3. INRS-Eau, Quebec.

Brees, M (2004) Corn Silage Budgets for Northern, Central and Southwest Missouri. http://www.agebb.missouri.edu/mgt/budget/fbm-0201.pdf (9/1/04).

Brown, J (2005) Ethanol Isn't Worth the Energy. On Line Opinion. Australia's e-Journal of Social and Political Debate. www.onlineopinion.com.au/view.asp?article=3859 (12/2/05).

Brown, LR (1997) The Agricultural Link: How Environmental Deterioration Could Disrupt Economic Progress. Worldwatch Institute, Washington, DC.

CalGasoline (2002) Ethanol is Not a Suitable Replacement for MTBE. www.calgasoline.com/factetha.htm (9/17/2002).

CIA (2005) Brazil Natural Gas. The Library of Congress Country Studies; CIA World Factbook. http://www.photius.com/countries/brazil/economy/brazil_economy_natural_gas.html. (9/5/05).

Coelho, ST, Bolognini, MF, Silva, OC, Paletta, CEM (2002) Biofuels in Brazil: The Current Situation. CENBIO – The National Reference Center on Biomass. Technical Texts. http://www.cenbio.org.br/in/index.html (11/12/2002).

Croysdale, D (2001) Belatedly, DNR Concedes Our Air is Clean. Daily Reporter Nov 6, 2001. http://www.dailyreporter.com/editorials/eds/nov07/asa11-7.shtml. (9/17/2002).

DOE (2002) Review of Transport Issues and Comparison of Infrastructure Costs for a Renewable Fuels Standard. U.S. Department of Energy, Washington, DC. http://tonto.eia.doe.gov/FTPROOT/service/question3.pdf (10/8/2002).

DOE (2005) Energy Efficiency and Renewable Energy. U.S. Department of Energy. Washington DC. http://www.eere.energy.gov/biomasss/eethanol.html. (1/6/06).

EPA (2002) More Pollution Than They Said: Ethanol Plants Releasing Toxins. New York Times, May 3, 2002.
ERAB (1980) Gasohol. Energy Research Advisory Board, U.S. Department of Energy, Washington, DC.
ERAB (1981) Biomass Energy. Energy Research Advisory Board, U.S. Department of Energy, Washington, DC.
FAO (2001) Food Balance Sheets. Food and Agriculture Organization of the United Nations, Rome.
Farrell, AE, Plevin, RJ, Turner, BT, Jones, AD, O'Hare, MO, Kammen, DM (2006) Ethanol can contribute to energy and environmental goals. Science 311:506–508.
Ferguson, ARB (2003) Implications of the USDA 2002 update on ethanol from corn. The Optimum Population Trust, Manchester, UK 3(1):11–15.
Ferguson, ARB (2004) Further implications concerning ethanol from corn. Draft manuscript for the Optimum Population Trust.
Giampietro, M, Ulgiati, S, Pimentel, D (1997) Feasibility of large-scale biofuel production. BioScience 47:587–600.
Hassett, KA (2006) National Review Magazine. http://en.wikipedia.org/wiki/American_Enterprise_Institute. (2/16/06).
Ho, SP (1989) Global warming impact of ethanol versus gasoline. Presented at 1989 National Conference. Clean Air Issues and American Motor Fuel Business. Oct 1989. Washington, DC.
Hodge, C (2002) Ethanol use in US gasoline should be banned, not expanded. Oil Gas J Sept 9:20–30.
Hodge, C (2003) More evidence mounts for banning, not expanding, use of ethanol in gasoline. Oil Gas J Oct 6:20–25.
Hodge, C (2005) Government and Fuels. Increased Air Pollution with the Consumption of Ethanol in Gasoline. http://www.arb.ca.gov/fuels/gasoline/meeting/2005/0502052ndopi (10/0/05).
Hoffman, TR, Warnock, WD, Hinman, HR (1994) Crop Enterprise Budgets, Timothy-Legume and Alfalfa Hay, Sudan Grass, Sweet Corn and Spring Wheat under Rill Irrigation, Kittitas County, Washington. Farm Bus Rep EB 1173, Washington State University, Pullman.
Illinois Corn (2004) Ethanol's Energy Balance. http://www.ilcorn.org/Ethanol/Ethan_Studies/Ethan_Energy_Bal/ethan_energy_bal.html (8/10/04).
Kim, Y (2002) World exotic diseases. In: Pimentel D (ed) Biological Invasions: Economic and Environmental Costs of Alien Plant, Animal, and Microbe Species. CRC Press, Boca Raton, FL, pp. 331–354.
Kuby, WR, Markoja, R, Nackford, S (1984) Testing and Evaluation of On-Farm Alcohol Production Facilities. Acures Corporation. Industrial Environmental Research Laboratory, Office of Research and Development, U.S. Environmental Protection Agency, Cincinnati, OH.
Larsen, K, Thompson, D, Harn, A (2002) Limited and Full Irrigation Comparison for Corn and Grain Sorghum. http://www.colostate.edu/Depts/SoilCrop/extension/Newsletters/2003/Drought/sorghum.html (9/2/2002).
Larson, WE, Cardwell, VB (1999) History of U.S. Corn Production. http://citv.unl.edu/cornpro/html/history/history.html (9/2/04).
Lieberman, B (2002) The Ethanol Mistake: One Bad Mandate Replaced by Another. Competitive Enterprise Institute. http://www.nationalreview.com/comment/comment-lieberman031202.shtml (9/17/2002).

Maiorella, B (1985) Ethanol. In: Blanch HW, Drew S, Wang DIC (eds) Comprehensive Biotechnology, vol 3. Pergamon Press, New York.
McCain, J (2003) Statement of Senator McCain on the Energy Bill. Press Release, Nov 19, 2003.
NAS (2003) Frontiers in Agricultural Research: Food, Health, Environment, and Communities. National Academy of Sciences, Washington, DC: http://dels.nas.edu/rpt_briefs/frontiers_in_ag_final%20for%20print.pdf (11/05/04)
NASS (2003) National Agricultural Statistics Service. http://usda.mannlib.cornell.edu (10/9/05).
National Center for Policy Analysis (2002) Ethanol Subsidies. Idea House. National Center for Policy Analysis. http://www.ncpa.org/pd/ag/ag6.html (9/09/2002).
Newton, PW (2001) Human Settlements Theme Report. Australian State of the Environment Report 2001. http://.deh.gov.au/soe/2001/settlements/settlements02-5c.html (10/6/05).
Niven, R (2005) UNSW Academic Criticizes Decision on Ethanol in Petrol. http://www.unsw.edu./news/pad/articles/2005/sep/Ethanol.html (10/9/05).
NPRA (2002) NPRA Opposes Ethanol Mandate; Asks Congress Not to Hinder Efforts to Maintain Supply. National Petrochemical & Refiners Association, Washington, DC: http://www.npradc.org/news/releases/detail.cfm?docid=164&archive=1 (9/17/2002).
Patzek, T (2004) Thermodynamics of the corn-ethanol biofuel cycle. Crit Rev Plant Sci 23(6):519–567.
Pimentel, D (1991) Ethanol fuels: Energy security, economics, and the environment. J Agric Environ Ethics, v. 4:1–13.
Pimentel, D (1998) Energy and dollar costs of ethanol production with corn. Hubbert Center Newsletter #98/2 M. King Hubbert Center for Petroleum Supply Studies. Colorado School of Mines, Golden, CO, 7 pp.
Pimentel, D (2001) The limitations of biomass energy. In: Meyers R (ed) Encyclopedia of Physical Science and Technology. 3rd ed., Vol. 2, Academic Press, San Diego, p. 159–171.
Pimentel, D (2003) Ethanol fuels: energy balance, economics, and environmental impacts are negative. Nat Resour Res, 12 (2): 127–134.
Pimentel, D, Patzek. T (2005) Ethanol production using corn, switchgrass, and wood and biodiesel production using soybean and sunflower. Nat Resour Res 14(1):65–76.
Pimentel, D, Pimentel, M (1996) Food, Energy and Society. Colorado University Press, Boulder, CO.
Pimentel, D, Harvey, C, Resosudarmo, P, Sinclair, K, Kurz, D, McNair, M, Crist, S, Sphritz, L, Fitton, L, Saffouri, R, Blair, R (1995) Environmental and economic costs of soil erosion and conservation benefits. Science 276:1117–1123.
Pimentel, D, Doughty, R, Carothers, C, Lamberson, S, Bora, N, Lee, K (2002) Energy inputs in crop production: comparison of developed and developing countries. In: Lal R, Hansen D, Uphoff N, Slack S (eds) Food Security & Environmental Quality in the Developing World. CRC Press, Boca Raton, FL, pp. 129–151.
Pimentel, D, Pleasant, A, Barron, J, Gaudioso, J, Pollock, N, Chae, E, Kim, Y, Lassiter, A, Schiavoni, C, Jackson, A, Lee, M, Eaton, A (2004a) U.S. energy conservation and efficiency: benefits and costs. Environ Dev Sustain 6:279–305.

Pimentel, D, Berger, B, Filberto, D, Newton, M, Wolfe, B, Karabinakis, E, Clark, S, Poon, E, Abbett, E, Nandagopal, S (2004b) Water resources: current and future issues. BioScience 54(10):909–918.

Pimentel, D, Cooperstein, S, Randell, H, Filiberto, D, Sorrentino, S, Kaye, B, Nicklin, C, Yagi, J, Brian, J, O'Hern, J, Habas, A, Weinstein, C (2006) Ecology of increasing diseases: population growth and environmental degradation. Environ Dev Sustain (in press).

PRB (2005) World Population Data Sheet. Population Reference Bureau, Washington, DC.

Schneider, SH, Rosencranz, A, Niles, JO (2002) Climate Change Policy Change. Island Press, Washington, DC.

Shapouri, H, Duffield, JA, Wang, M (2002) The Energy Balance of Corn Ethanol: An Update. Report 813. USDA, Office of Energy Policy and New Uses, Agricultural Economics, Washington, DC.

Shapouri, H, Duffield, J, McAloon, A, Wang, M (2004) The 2001 Net Energy Balance of Corn-Ethanol (Preliminary). U.S. Department of Agriculture, Washington, DC.

Spirits Low (1999) Spirits Low as Brazil Alcohol Car in Trouble Anew. Reuters Ltd., Nov 22, 1999. http://www.climateark.org/articles/1999/alcocaro.htm (11/15/04).

Stanton, TL (1999) Feed Composition for Cattle and Sheep. Report 1.615. Cooperative Extension, Colorado State University, Boulder, CO.

Transportation Research Board (2006) Integrating Sustainability into the Transportation Planning Process, 40. http://www.trp.org/publications.conf/CP37.pdf. (2/16/06).

USCB (2004–2005) Statistical Abstract of the United States 2004–2005. U.S. Census Bureau, U.S. Government Printing Office, Washington, DC.

USDA (1991) Corn-State. Costs of Production. Report 94018. U.S. Department of Agriculture, Economic Research Service, Economics and Statistics System, Washington, DC.

USDA (1997a) Farm and Ranch Irrigation Survey (1998). 1997 Census of Agriculture, vol 3. Special studies, part 1. USDA, Washington, DC.

USDA (1997b) 1997 Census of agriculture. U.S. Department of Agriculture. http://www.ncfap.org. (8/28/2002).

USDA (2002) Agricultural Statistics. U.S. Government Printing Office, Washington, DC.

USDA (2003) Agricultural Statistics. USDA. I – 1 – XV-34 p. U.S. Government Printing Office, Washington, DC.

Wald, MI (2006) Corn farmers smile as ethanol prices rise, but experts on food supplies worry. New York Times (National). A13.

Wereko-Brobby, C, Hagan, EB (1996) Biomass Conversion and Technology. Wiley, Chichester, UK.

WHO (2000) Nutrition for health and development: a global agenda for combating malnutrition. World Health Organization. http://www.who.int/nut/documents/nhd_mip_2000.pdf (11/3/04).

Wilcke, B, Chaplin, J (2000) Fuel Saving Ideas for Farmers.Minnesota/ Wisconsin Engineering Notes. http://www.bae.umn.edu/extens/ennotes/enspr00/fuelsaving.htm (9/2/04).

World Environment News (2006) Ethanol Could Reduce Fossil Fuel Need. http://www.planetrak.com/dailynews story.cfm/34678/story.htm. (1/27/06).

Worldwatch Institute (2001) Vital Signs. Norton, New York.
Youngquist, W (1997) GeoDestinies: The Inevitable Control of Earth Resources Over Nations and Individuals. National Book Company, Portland, OR.
Youngquist, W, Duncan RC (2003) North American natural gas: data show supply problems. Nat Resour Res 12(4):229–240.

Manuscript received January 25; accepted February 27, 2004.

Rev Environ Contam Toxicol 189:43–87

Arsenic Behaviour from Groundwater and Soil to Crops: Impacts on Agriculture and Food Safety

Alex Heikens, Golam M. Panaullah, and Andy A. Meharg

Contents

I. Introduction 43
II. Arsenic Chemistry and Behaviour 44
A. Soil 44
B. Crops 48
C. Foods 60
III. Situation in Bangladesh 65
A. Agriculture 65
B. Foods 71
C. Management Options 77
IV. Future Requirements 78
A. Agricultural Sustainability 78
B. Food Safety 79
C. Increasing Quantity and Quality of Chemical Analyses 79
V. Conclusions 80
Summary 80
Acknowledgments 81
References 81

I. Introduction

Arsenic (As) in groundwater poses a major health concern in Asia. To date, unacceptably high As levels in groundwater resources have been found in parts of Bangladesh, Cambodia, China, India, the Lao People's Democratic Republic, Mongolia, Myanmar, Nepal, Pakistan, Thailand, and Viet Nam

---

Communicated by George W. Ware.

A. Heikens (✉)
Food and Agriculture Organization of the United Nations, Regional Office for Asia and the Pacific (FAO-RAP), 39 Phra Atit Road, Bangkok 10200, Thailand.

G.M. Panaullah
International Maize and Wheat Improvement Center (CIMMYT), Office in Bangladesh, P.O. Box 6057, Gulshan, Dhaka 1212, Bangladesh.

A.A. Meharg
School of Biological Sciences, University of Aberdeen, Aberdeen AB24 3UU, United Kingdom.

The views expressed in this paper are those of the authors and do not necessarily reflect the views of FAO, CIMMYT, or the University of Aberdeen.

(Mandal and Suzuki 2002; Ng et al. 2003). Numerous reviews on arsenic in the environment have been published recently (Chakraborti et al. 2002; Mandal and Suzuki 2002; Ng et al. 2003; WHO 2001). Bangladesh was reported to have the highest percentage (~25%) of contaminated shallow tube wells (STW), and an estimated 30 million people are dependent on these water wells for domestic purposes.

In Bangladesh, groundwater supplies not only domestic drinking water but water for irrigated agriculture as well. In the past three decades, the number of STW used for irrigation purposes has increased dramatically in Bangladesh, and the dry season rice production ("Boro rice") depends heavily on STW. The introduction and widespread use of STW have been one of the pillars of Bangladesh's "Green Revolution," resulting in the country's self-sufficiency in its staple crop (rice) production today.

The high levels of As in groundwater in the Bengal delta are of geogenic origin. The two main theories on the underlying mechanism are oxidation of As-containing pyrite and the reductive dissolution of iron (hydr)oxides (FeOOH) in combination with the release of As (Ahmed et al. 2004; McArthur et al. 2004; Smedley and Kinniburgh 2002; Zheng et al. 2004). More and more evidence in favour of the second mechanism has been obtained in the last few years (Ravenscroft et al. 2001). Although oxidation of pyrite due to a lowering water table seems to be excluded, there still is controversy whether groundwater extraction for irrigation influences As behaviour in the shallow aquifer. It has been suggested that it may cause an increase of infiltration of organic matter (OM) into the shallow aquifer, which facilitates reductive dissolution of FeOOH (Harvey et al. 2002).

Dependent on many factors, As in irrigation water may negatively affect crop production and food safety (food chain contamination) (Brammer 2005; Duxbury and Zavala 2005), and the use of As-rich irrigation water could result in elevated soil concentrations. If reflected by the crops including the edible parts, this may pose a human health risk. Eventually, continual accumulation of As in the soil over time could lead to phytotoxic soil concentrations, leading to reduction in yields and other negative impacts.

The aims of this review are (1) to provide an overview of current knowledge related to As risks in irrigation water and its potential impacts on food safety and crop production, (2) to summarize specific knowledge on As in agriculture, and (3) to identify needs for future activities in Bangladesh. The focus is on rice, as it is the staple crop in Bangladesh, as well as As sensitivity of rice grown under flooded conditions, and the high input of irrigation water into the paddy fields.

II. Arsenic Chemistry and Behaviour

A. Soil

The element As is ubiquitously present in soils, normally at trace quantities (Matschullat 2000). Worldwide, natural soil concentrations are around

5 mg/kg, but this can vary depending on the origin of the soil (Mandal and Suzuki 2002). To date, most research has focused on the behaviour of As under nonflooded (aerobic) soil conditions, which is relevant for many crops. Rice, being the most widely produced crop in Bangladesh, is cultivated under flooded (anaerobic) conditions. Redox conditions have a great impact on As behaviour in soil, emphasising the need of understanding As behaviour under anaerobic conditions as well.

As Speciation in the Soil. Both organic and inorganic forms of As exist in the environment. The most important inorganic forms are arsenate, AsV ($H_2AsO_4^-$ at pH < 7.0; $HAsO_4^{2-}$ at pH > 7.0), and arsenite, AsIII, ($H_3AsO_3^0$ at pH < 9.2) (Fitz and Wenzel 2002). Monomethylarsenic acid, or MMA [$CH_3AsO(OH)_2$], and dimethylarsenic acid, or DMA [$(CH_3)_2AsO(OH)$], are the most common organic species in the soil, but their natural presence is low compared to inorganic As. Usually only low quantities of MMA and DMA are present with the highest levels found in flooded paddy fields (Abedin et al. 2002c; Fitz and Wenzel 2002).

Redox Conditions. As speciation in the soil is largely redox controlled. Under aerobic conditions (oxidizing environment), AsV is predominant, whereas AsIII predominates under anaerobic conditions (reducing environment) (Fitz and Wenzel 2002; Takahashi et al. 2004). It was reported, in an experimental paddy field, that under nonflooded conditions 30% of the As was present as AsIII and up to 70% as AsIII under flooded conditions (Takahashi et al. 2004). During the nonflooded period, the Eh value was constantly high (~600 mV) and decreased continuously during the flooded period down to approximately 0 mV at 0.1–0.2 m depth (Takahashi et al. 2004). Masscheleyn et al. (1991) reported that under oxidizing conditions (Eh 500 to 200 mV), As was mainly present as AsV (>95% of the total soluble As) and that the solubility was relatively low. With lower Eh values (0 to −200 mV), AsIII became by far the predominant species and the solubility of As increased sharply (Masscheleyn et al. 1991). The importance of redox is strongly related to the behaviour of FeOOH.

Role of Fe Hydroxides. Fe redox chemistry is the predominant factor in regulating As behaviour in the soil. FeOOH is the main sorbent for both AsV and AsIII and (Fitz and Wenzel 2002; Norra et al. 2005; Takahashi et al. 2004). AsV is most strongly adsorbed by FeOOH, explaining why the distribution coefficient for AsV is greater than for AsIII (Lambkin and Alloway 2003; Takahashi et al. 2004). The adsorption isotherm for AsV is highly nonlinear (Mahimairaja et al. 2005). In batch and column experiments, it was found that AsV is rapidly adsorbed with most of the added AsV (~90%) adsorbed within 48 hr (Williams et al. 2003). In a water–sediment system, a great amount of As is sorbed by the sediment from the overlying water, even at low As concentrations in water (Mahimairaja et al. 2005). For example, Norra et al. (2005) found that after flooding a paddy field with water containing 0.519 mg/L As, overnight standing of the water resulted in an As concentration of 0.016 mg/L.

FeOOH is mainly present in the clay size soil fraction (<2 μm), and clayey soils therefore generally have a higher As content compared to more sandy soils (Fitz and Wenzel 2002; Mahimairaja et al. 2005). At the same total soil concentration, clayey soils are less toxic compared to sandy soils because As is more strongly bound in the clayey soils. Under specific soil conditions, other sorption substrates can also be relevant. For example, in calcareous soils, carbonate minerals are important, whereas in acidic soils manganese oxides are important (Mahimairaja et al. 2005).

Under anaerobic conditions, FeOOH is readily dissolved, releasing mainly AsV into the soil solution that is then reduced to AsIII (Fitz and Wenzel 2002). Reduction of AsV to AsIII before desorption from FeOOH has also been reported (Takahashi et al. 2004), who proposed that desorption of As from FeOOH was responsible for the high As concentrations in the soil water (Takahashi et al. 2004). Masscheleyn et al. (1991) described the release of As into the soil solution as a two-step process. After 3 d flooding, reduction and desorption of some As from FeOOH was observed without any dissolution of FeOOH. After 35 d, a sharp increase in As and Fe was observed, suggesting dissolution of FeOOH and release of As. Because AsV was present in the soil water, it seems that AsV was first released from FeOOH and then slowly reduced to AsIII in the soil water. In contrast with Takahashi et al. (2004), Masscheleyn et al. (1991) showed that although As desorption took place before reductive dissolution of FeOOH, the latter contributed most to the final As concentrations in the soil water. These contradicting results might be explained by the process of aging, by which chemicals in the soil become less available over time. For As, it has been proposed that As penetrates deeper into the FeOOH, making it less sensitive to dissolution under reducing conditions (Zobrist et al. 2000). Under aerobic conditions FeOOH is relatively insoluble and serves as a sink for As. Considering the cycles of aerobic and anaerobic soil conditions in paddy fields, Fe and As chemistry are of a dynamic nature, making it less likely that aging plays a role in paddy fields. Although AsIII predominates in the bulk soil under flooded conditions, it is unclear whether an oxidized root zone will influence As speciation before uptake.

Takahashi et al. (2004) reported that As and Fe concentrations in irrigation water were higher compared to the soil water concentrations during the nonflooded period, suggesting that part of these elements are being adsorbed by the soil. Under flooded conditions, soil water concentrations increased sharply in the first meter of the profile (especially at 0.2–0.5 m) and were higher than the irrigation water concentrations. This observation is in agreement with Duxbury and Loeppert, who found that As pore-water concentrations increase with the duration of flooding and that these can largely exceed irrigation water concentrations, reaching levels above 1 mg/L (personal communication, Duxbury and Loeppert 2004).

Phosphate. Phosphate (PO_4^{3-}) is an analogue of AsV, making it an important factor in the behaviour of As in aerobic soils (Lambkin and Alloway 2003; Mahimairaja et al. 2005; Williams et al. 2003). Both ions compete for sorption sites on soil surfaces and for uptake by plants. Addition of PO_4^{3-} to aerobic soils has been suggested to increase the mobility and bioavailability of AsV. However, there is controversy whether this will enhance the uptake of As because PO_4^{3-} also competes with AsV for uptake. Lambkin and Alloway (2003) suggested that input of As may cause leaching of PO_4^{3-} from the topsoil, which could affect crop production. PO_4^{3-} is not an analogue of AsIII, making the presence of PO_4^{3-} in anaerobic soils probably less relevant in terms of As behaviour (Takahashi et al. 2004). It cannot be excluded that PO_4^{3-} does influence As behaviour in the rhizosphere, where aerobic conditions can occur under flooded conditions. Other ions may also influence As adsorption, but the impact seems to be less compared to PO_4^{3-} (Cornu et al. 2003; Mahimairaja et al. 2005; Williams et al. 2003).

pH. AsV adsorption decreases with increasing pH, in particular above pH 8.5, while the opposite occurs for AsIII. The adsorption maximum for AsV on FeOOH lies around pH 4, while for AsIII the maximum is found at approximately pH 7–8.5 (Fitz and Wenzel 2002; Mahimairaja et al. 2005; Masscheleyn et al. 1991). This result is in agreement with Williams et al. (2003), who reported that in the pH range 7–10, the adsorption of AsV (present as an anion) decreased approximately 50%; this was related to the variable charged surfaces of soil surfaces such as Fe oxides, which become more negatively charged above pH 7. Below this value, pH did have not much influence on AsV adsorption. The relationship may be more complex. Cornu et al. (2003) reported that the influence of pH on AsV sorption depended on the electrolyte used and the presence of humic acids (HA). In a Na electrolyte, the AsV sorption was pH dependent (negative correlation) and sorption was inhibited in the presence of HA. In a Ca electrolyte, pH had no influence on As sorption in the absence of HA and only had a minor effect of sorption in the presence of HA.

Organic Matter. Organic compounds can compete with As for sorption sites: HA has been shown to reduce As sorption by binding As, and dissolved OM can also reduce sorption of As to soil particles through redox reactions and by the formation of soluble complexes with As (Fitz and Wenzel 2002; Mahimairaja et al. 2005). According to Fitz and Wenzel (2002), there is no evidence available proving that OM is an important As sorbent.

Microorganisms. Microbial activity can influence As behaviour via various mechanisms such as redox reactions with Fe and As and via (de)methylation of As species. Primary producers such as algae reduce AsV to AsIII before methylation to MMA and DMA, which are then excreted (Fitz and Wenzel 2002; Mahimairaja et al. 2005). Microbial activity is closely involved in the reductive dissolution of FeOOH in the presence of organic matter, followed by the release and reduction of AsV (Islam et al. 2004a). Zobrist et al. (2000) showed that certain bacteria can reduce both AsV and FeIII and that the reduction of AsV, and its dissolution, can take place without the dissolution of the sorbent, which was ferrihydrite in their experiments. Masscheleyn et al. (1991) suggested that microbial mineralization of OM, using AsV as an electron acceptor is the main reason for reduction of As V to AsIII (Masscheleyn et al. 1991). The effect of microbial activity on As behaviour is likely to be of particular importance in the rhizosphere.

Volatilization. Formation of gaseous As has been mentioned by various authors (Abedin et al. 2002c; Mahimairaja et al. 2005). As summarized by WHO (2001), volatilization can substantially contribute to As removal from topsoils, up to 12%–35% yr. No specific data for anaerobic soils such as flooded paddy fields have been found, making it yet not possible to assess the importance for As behaviour in paddy fields.

B. Crops

Rhizosphere. Soil chemistry often focuses on bulk soil analysis, but conditions in the microenvironment around the roots (the rhizosphere) may deviate substantially from the bulk soil. As summarized by Fitz and Wenzel (2002), plants influence pore-water composition by uptake of substances (e.g., essential and nonessential elements) and excretion of substances such as organic acids, H^+, O_2, and CO_2. Microorganisms in the rhizosphere also influence its composition. The pH in the rhizosphere can differ up to a factor of two compared to the bulk soil pH. The source of nitrogen plays an important role because nitrogen uptake has the greatest influence on the cation/anion uptake ratio. NH_4^+ uptake can reduce rhizosphere pH whereas NO_3^- uptake can have the opposite effect. NO_3^- is a strong oxidizing agent, and application to the soil may inhibit As solubility by the oxidation of AsIII to AsV and the subsequent adsorption on FeOOH (Harvey et al. 2002; Nicolas et al. 2003). The buffering capacity of the soil will also influence the rhizosphere pH. Because As, Fe, and PO_4^{3-} behaviour in the soil are closely related to each other, it can be expected that plant processes related to Fe and PO_4^{3-} uptake may also influence As bioavailability and uptake.

Fe Plaque Under Anaerobic Soil Conditions. Under anaerobic conditions, rice plants can transport oxygen from the leaves to the roots, resulting in a leakage of O_2 to the rhizosphere. This transfer can create aerobic conditions in the rhizosphere and can cause precipitation of FeOOH on the roots, also known as Fe plaque. Microbial oxidation processes may also contribute to the formation of Fe plaque, which may influence As bioavailability and uptake (Fitz and Wenzel 2002; Meharg 2004).

In a study on wetland plants, Weiss et al. (2004) found that the release of oxygen by roots showed diurnal and seasonal variation. Therefore, a cycle of Fe precipitation under aerobic conditions and Fe dissolution under anaerobic conditions can occur at the root system. The rhizosphere contained much higher poor crystalline Fe (66% ± 7%) than in the nonrhizosphere soil (23% ± 7%). The percentage of Fe(III)-reducing bacteria (FeRB) was significantly correlated with poor crystalline Fe (Weiss et al. 2004). Both Fe(II)-oxidizing bacteria (FeOB) and FeRB have been found at the same places at the root system (Weiss et al. 2003). Weiss et al. (2004) confirmed the hypothesis that Fe(III) minerals are more reactive in the rhizosphere than in the bulk soil, because both chemical and microbial reduction of Fe in the rhizosphere was faster and occurred to a greater extent compared to the nonrhizosphere soil. They also proposed that wetland vegetation was not only causing accumulation of poorly crystalline Fe in the active rhizosphere, but that much of it in the nonrhizosphere soil originated from senesced roots. The process of reductive dissolution can nearly or completely dissolve the pool of amorphous Fe(III) oxides. It was estimated that ~80% of the Fe plaque was involved in the Fe cycling in the rhizosphere (Weiss et al. 2004). Labile OM will facilitate reductive dissolution of FeIII by FeRB in the rhizosphere, and it is likely that the rhizosphere contains more labile OM than does the nonrhizosphere soil (Weiss et al. 2003, 2004).

Fe plaque has been commonly observed in roots of wetland plants including rice, and it may play an important role in As tolerance (Meharg 2004). Currently, the data are contradictory. Liu et al. (2004) showed in hydroponic experiments the potential importance of Fe-plaque formation induced by P deficiency on AsV uptake by rice plants. In their experiments, 24 hr after transplantation from a P-sufficient solution to a P-deficient solution, Fe plaque formation was observed, containing high levels of AsV. Under P-deficient conditions, the translocation of AsV from the roots to shoots was reduced while the AsV concentrations in the roots were increased. The rice cultivar-dependent level of Fe-plaque formation was negatively correlated to the AsV levels in shoots. The authors concluded that the influence of Fe-plaque formation on As uptake and translocation should be further investigated considering its suggested capability to limit As translocation to aboveground parts of plants. A shortcoming was that only AsV was taken into account whereas AsIII is dominant in the bulk paddy soil (Fitz and Wenzel 2002).

Based on the findings of Liu et al. (2004), Chen et al. (2005) studied the effect of Fe plaque and PO_4^{3-} on the uptake kinetics of AsIII and AsV with excised roots of rice seedlings. In the absence of Fe plaque, PO_4^{3-} in the solution inhibited AsV uptake, which can be explained by competition between AsV and PO_4^{3-} for uptake. In the presence of Fe plaque, the effect of PO_4^{3-} was reduced, probably because PO_4^{3-} is then also competing with AsV for sorption sites at Fe plaque. For AsIII, both Fe plaque and PO_4^{3-} did not have any significant effect on AsIII uptake, which is in agreement with the fact that PO_4^{3-} is not competing with AsIII for uptake (Abedin et al. 2002c). AsIII uptake seemed to be stimulated in the presence of Fe plaque, for which no possible explanation was found.

Uptake. AsIII and AsV are taken up by different mechanisms that are not related to each other. AsV is taken up via the high-affinity phosphate uptake system (Meharg 2004). In two similar papers, no effect of PO_4^{3-} application on AsV uptake was found when rice was grown in pots with soil irrigated with AsV-contaminated water (Abedin et al. 2002a,b). Abedin et al. (2002a) suggested that the plants were effectively exposed to AsIII and not to AsV because of the reducing soil conditions. An alternative explanation is that PO_4^{3-} competes with AsV for both sorption at Fe plaque and uptake, minimizing the effect of PO_4^{3-} on AsV uptake (Chen et al. 2005). Overall, there are no indications that PO_4^{3-} plays an important role in As uptake by rice under flooded field conditions. As summarized by several authors (Abedin et al. 2002c; Fitz and Wenzel 2002), the effect of PO_4^{3-} additions to As-contaminated soils to minimize As uptake is also controversial under nonflooded conditions. Most studies showing inhibition of AsV uptake by PO_4^{3-} were performed in hydroponics, excluding the competition between PO_4^{3-} and AsV for sorption places at the soil surface.

AsIII is a small uncharged molecule and is actively taken up by water channels (aquaporins) in the roots (Meharg and Jardine 2003). Aquaporins are known to transport various substances such as water and glycerol. Meharg and Jardine (2003) proved the uptake of AsIII in rice via aquaporins by competition experiments with glycerol and antimonite, the latter being an analogue of AsIII. AsV uptake did not show any response to glycerol and antimonite. The knowledge on AsIII uptake by rice has evolved only recently. Abedin et al. (2002c) performed experiments to determine short-term uptake kinetics of four As species (AsIII, AsV, MMA, DMA) in 7-d-old seedlings of eight rice cultivars from Bangladesh. The Boro (dry season) cultivars took up less AsIII and AsV than Aman (rainy season) cultivars, which may be related to physiological or morphological differences between the root systems of Boro and Aman cultivars (Abedin et al. 2002c). This observation does not automatically imply that under field conditions Boro accumulates less As than Aman, because Boro is irrigated with groundwater that can contain As whereas Aman is rain fed. Uptake of AsIII was active, at approximately the same rate as AsV, whereas the uptake of DMA and MMA was much lower than AsIII and AsV. Their hypothesis that aquaporins were the route of AsIII uptake was later confirmed by Meharg and Jardine (2003). Uptake kinetics characteristics are suggested to be of importance to select rice cultivars for areas with As contamination (Abedin et al. 2002c). It is hypothesized that As(III) will dominate influx in rice grown under flooded conditions, despite the oxidized rhizosphere. AsIII uptake should also be considered under aerobic soil conditions, because microanaerobic spots will be present in such soils (Meharg and Jardine 2003).

The uptake mechanism of organic As is largely unclear (Meharg 2004). It seems that MMA and DMA are taken up by rice plants, but the rate of uptake is much lower compared to inorganic As (Abedin et al. 2002c). It is still unknown via what pathway organic As is taken up.

To date, it has not been possible to quantify the relationship between As in soil and plants. Most papers only include total As concentrations in the soil and the As concentration in the irrigation water. The ability to predict As bioavailability in the soil and uptake by plants would be of major importance to evaluate current and future soil concentrations.

Translocation and Accumulation. With the exception of hyperaccumulators, translocation of inorganic As from the roots to the aboveground parts is generally limited. Organic arsenic is more readily translocated but the uptake is much lower compared to inorganic As. For example, Carbonell et al. (1998) and Carbonell-Barrachina et al. (1998) found that wetland marsh grasses mainly translocated DMA to the shoots whereas the other As species were largely retained in the roots. The DMA concentrations in roots were however much lower than other As species, resulting in comparable concentrations of all species in the shoots. When exposed to AsIII, tomato plants accumulated most of the As in the roots whereas bean plants accumulated As mainly in the stems and leaves (Carbonell-Barrachina et al. 1997). For turnip plants, accumulation of As species was ranked as MMA < DMA < AsIII < AsV (Carbonell-Barrachina et al. 1999). In their pot experiments with rice plants exposed to As added via AsV in irrigation water, Abedin et al. (2002a,b) ranked plant parts according to the As concentrations as follows: root > straw > husk > grain. Concentrations in all plant parts increased with the exposure concentration, a common observation for other plants as well (Bleeker et al. 2003; Carbonell et al. 1998; Carbonell-Barrachina et al. 1997, 1998; Hartley-Whitaker et al. 2001; Sneller et al. 1999b). The ranking of plant parts according to As accumulation is regularly used as "evidence" that aboveground edible parts are no risk to human health. It is, however, the absolute concentration of inorganic As in the edible parts that should be evaluated, regardless of As concentrations in other parts of the plant.

A number of plants such as various ferns have been identified as hyperaccumulators. These plants actively take up As and store it in the aboveground parts (Fitz and Wenzel 2002). The potential of such plants to remediate contaminated locations is currently under investigation. One of the limitations is the long time span needed to remove As to acceptable levels, making it most likely unsuitable for agricultural soils.

Metabolism. After uptake of AsV, it is rapidly reduced to AsIII, causing oxidative stress; this induces the formation of antioxidants such as glutathione, which also is the precursor of phytochelatins (PC). The latter are known to detoxify heavy metals such as cadmium, and the same has been reported for As (Meharg and Hartley-Whitaker 2002; Sneller et al. 1999a, 2000). In spite of the rapid reduction of AsV to AsIII, high levels of AsV have been reported in plant material. Abedin et al. (2002b) reported that more than 70% of the As in rice straw was present as AsV. A drawback of the extraction technique with trifluoroacetic acid (TFA) is that approximately 20% of AsV is reduced to AsIII, making TFA only suitable to determine the sum of inorganic As. It suggests that almost all As in the straw was present as AsV.

Oxidation of AsIII in plants may also take place. Schmidt et al. (2004) reported that in plants exposed to AsIII, shoots and leaves mainly contained AsIII but also some AsV. AsV found in plant extracts may have originated from AsIII–PC complexes when extraction solutions with $pH > 7.2$ are used (Meharg and Hartley-Whitaker 2002). In their research on As–PC complexes, Raab et al. (2005) reported 14 different As species in roots of sunflowers (*Helianthus annuus*) grown on Perlite and exposed to AsIII or AsV up to 5 d after an initial growth period of 67 d.

Many methylated As species have been found in plants as well, but only in minor amounts (Dembitsky and Rezanka 2003). It is unclear whether organic As species found in plants were taken up from the soil or were formed by the plants. Sneller et al. (1999b) mentioned that methylation of AsV by plants can take place under P-deficient conditions. In contrast, according to Meharg and Hartley-Whitaker (2002) methylation in plants has not been proved yet as the correct experiments have not been conducted. They do recognize that mechanisms potentially involved in As methylation are present in plants and if it occurs this may have significant implications on As tolerance, considering the differences in toxicity between As species. All experiments so far have added As to soil or water, after which plants were grown. In most cases, As alteration of speciation in the growth matrix was not checked. The few experiments that did so reported changes in As speciation. To prove methylation of inorganic As in plants, plants need to be given a short pulse of inorganic arsenic and then transferred to As-free media. Preferably, this should be done in a continuous-flow hydroponics system to rule out As excretion and reabsorption by the roots, with speciation followed over time.

Effects. Meharg and Hartley-Whitaker (2002) summarized the toxic effects of AsIII and AsV. After uptake of AsV, it is rapidly reduced to AsIII. Between AsV uptake and reduction to AsIII, AsV can compete with PO_4^{3-} by, e.g., replacing PO_4^{3-} in ATP, thereby disturbing the energy flow in the cell. A high PO_4^{3-} status is therefore likely to be needed to protect ATP (Meharg and Hartley-Whitaker 2002). AsIII reacts with sulfhydryl groups (–**SH**) groups of enzymes and tissue proteins, which can cause inhibition of cellular function and finally death. Exposure to As also influences concentrations of other elements in plant tissue (Williams et al. 2005). For example, Carbonell et al. (1998) found that P was reduced in the roots of rice plants when exposed to AsIII or AsV, while P in the shoots was increased.

As referred by Meharg and Hartley-Whitaker (2002), for example, a specific form of As toxicity to rice known as straighthead disease has been reported in the United States. This observation was related to rice production on former cotton fields heavily contaminated with MMA used as a herbicide. It was most frequently observed on sandy loam soils but seldom on clay soils. Affected plants were usually found in spots scattered throughout a field. Straighthead disease is a physiological disorder that causes panicle sterility. Visual symptoms are empty panicles standing upright instead of bending downward at maturity. In addition to As, a high OM content seemed to play a role as well. Rice cultivars show a great variation in their tolerance to MMA, which was used to select/develop tolerant cultivars. Other control measures include draining fields with a history of the disorder just before internode elongation.

A common ranking of plant parameters according to sensitivity to metals including As is as follows: root length > root mass > shoot length > total mass (root plus shoot) > shoot mass > germination (Abedin and Meharg 2002). This finding is in agreement with Abedin et al. (2002b), who found that root biomass production of rice plants was most sensitive to As whereas plant height was not very sensitive. Carbonell-Barrachina et al. (1998) reported for coastal marsh grasses that dry-matter production of roots was most sensitive. Abedin and Meharg (2002) proposed that shoot height can be used in the field as an indicator. However, Abedin et al. (2002a) reported that shoot height is much less sensitive than root length. They found that grain yield was the most sensitive parameter tested. Abedin and Meharg (2002) proposed the next chain of effects: reduced shoot height, reduced leaf area, reduced photosynthesis, reduced yield. It is likely that toxicity to the root system is actually the first step.

Relative Toxicity of As Species. Caution should be used with a generalized classification of As species according to toxicity, which may depend on experimental conditions and plant species. Hydroponic experiments may elucidate the intrinsic toxicity of an As species but do not show relevancy to the field situation. For the latter, it is needed to assess the exposure of plants under semifield conditions to identify the relevant As species. Taking that into account, inorganic As is generally regarded more toxic than organic As with AsIII as the most toxic form (Dembitsky and Rezanka 2003; Fitz and Wenzel 2002; Liu et al. 2004; Mahimairaja et al. 2005; Meharg and Hartley-Whitaker 2002). In contrast, Carbonell et al. (1998) and Carbonell-Barrachina et al. (1998) found in their experiments with marsh grasses grown hydroponically that organic As was more toxic than inorganic As, with MMA as the most toxic form and AsV as the least toxic form.

Carbonell-Barrachina et al. (1999) exposed turnip to MMA, DMA, AsV, and AsIII in hydroponics and found that organic As was more toxic than inorganic As and MMA was more toxic than DMA. Based on hydroponic experiments, Abedin and Meharg (2002) reported that AsIII was more toxic than AsV in terms of shoot height and germination of rice, but AsV was more toxic than AsIII in terms of root growth.

Tolerance. AsV tolerance is related to PO_4^{3-} metabolism by suppression of the high-affinity AsV/PO_4^{3-} plasma membrane cotransporters (Bleeker et al. 2003; Hartley-Whitaker et al. 2001; Meharg 2004; Meharg and Hartley-Whitaker 2002; Sneller et al. 1999b). Most AsV-tolerant plants accumulate less AsV than nontolerant plants, and this idea could be used to breed/select rice cultivars with a low accumulation of As. Plants not tolerant to AsV can be made more tolerant by increasing their PO_4^{3-} status (Meharg and Hartley-Whitaker 2002). In tolerant plants, the AsV/PO_4^{3-} system is usually continuously suppressed and is therefore insensitive to the plant PO_4^{3-} status. Meharg and Hartley-Whitaker (2002) concluded that "a decreased sensitivity in plants with a high PO_4^{3-} status does not result from a lower AsV influx but probably from a higher PO_4^{3-} level in the cytoplasm allowing PO_4^{3-} to compete more effectively with AsV for binding sites such as ATP. This could reduce the AsV toxicity in the cell."

Enhanced rates of PC accumulation may be an additional protection mechanism in AsV-hypertolerant plants. Bleeker et al. (2006) found that in AsV-hypertolerant common velvetgrass (*Holcus lanatus*) the enhanced PC-based sequestration was caused by an increase in AsV reductase activity, not by an enhanced PC synthesis capacity as such. They also reported that AsV tolerance was not correlated to AsIII tolerance.

Tolerance to AsIII is largely unknown. If there is any relevant variation in AsIII tolerance, this may be found in variation in glutathione levels [GSH forms complexes with As(III) and is a substrate for PC synthase and As transmembrane transporters] and/or the activity of certain ATP-binding cassette transporters that transport the formed complexes to the vacuole (personal communication. H. Schat, 2005).

Up to the present, studies on the genetics behind As tolerance have only focused on AsV. In wild grasses, a polymorphism in arsenic uptake and tolerance has been observed, and the tolerance is under single gene control. Recently, the same mechanism has been identified in rice (Dasgupta et al. 2004). If AsV uptake is relevant in rice grown under flooded conditions, this can be an important finding to develop/select rice cultivars with a high tolerance and a low As uptake. However, indications that AsIII predominates the As uptake suggest that suppression of the AsV/PO_4^{3-} system is less relevant to As uptake and tolerance in rice. The relative importance of As species under semifield conditions needs to be quantified to focus further research on the environmentally relevant species.

Toxicity Data. Basically three types of experimental setups are being used to evaluate As uptake and toxicity to plants. First, most experiments have been performed in hydroponics. Although these provide insight in differences in intrinsic toxicity between As species, uptake mechanisms, and translocation, such data cannot be extrapolated to the field because all interactions with the soil matrix are neglected. Second, experiments have been done with soils to which a certain amount of As was added before the experiment ("spiked soil"). Although the soil matrix is present in this setup, there are limitations to its use. The bioavailability of As added as a water-soluble salt immediately before growing the plants will probably be much higher compared to the same total soil concentration in the field where As is mainly sorbed to FeOOH. In none of the experiments that used this method was the contaminated soil aged for a substantial period before the plants were grown. It can be expected that As in the soil is not in steady state in such experiments and that the plants are effectively exposed to higher concentrations compared to a field situation with the same total As concentration (Duxbury and Zavala 2005). Third, others have added As via irrigation water to the soil during the experiments. Compared to the other two setups, this is most in agreement with the field situation in Bangladesh, for example. Although to a lesser extent than with spiked soils, this setup does neglect that As levels in irrigation water in the field are relatively constant and that As is slowly added to the soils over a period of many years. Ideally, experiments should be performed with field-contaminated soils and constant As concentrations in irrigation water.

To date, toxic effects have only been related to the irrigation water concentration or the total soil concentration. A direct relationship between one of these parameters and As uptake and toxicity is, however, unlikely. Dose–response relationships based on irrigation water concentrations or on total soil concentrations are therefore only valid for the experiment from which those were derived. With all these limitations in mind, a number of studies are summarized and discussed.

Hydroponics. Abedin and Meharg (2002) exposed eight Bangladesh rice varieties to AsIII and AsV and tested for germination and seedling growth. Exposure concentrations were 0, 0.5, 1.0, 2.0, 4.0, and 8.0 mg/L. Germination experiments were carried out on moist filter paper disks and germination was counted after 7 d. Germination was slightly inhibited at 0.5 and 1 mg/L. At 2 mg/L, inhibition was more than 10%. AsIII was more toxic than AsV. No significant difference between Boro and Aman cultivars in terms of germination was observed. Seedling growth experiments were carried in nutrient solutions with Ca, K, Mg, Na, NO_3, and SO_4. No PO_4^{3-} was added to avoid interaction with AsV. Five-day-old seedlings were exposed for 7 d. Root tolerance index (RTI, exposure length/control length of the longest root) and relative shoot height (RSH: exposure length/control length of the shoot) were used as measures of toxicity. Root growth was dependent on As species, rice variety, and season. In general, root growth was inhibited ~20% at 0.5 mg/L, and AsV was more toxic than AsIII. The latter may be explained by the absence of PO_4^{3-} in the growth solution. Boro cultivars were more tolerant than Aman cultivars, but the differences in tolerance were very small. The shoot height was also affected. At 0.5 mg/L, the shoot height was reduced ~30%. Although AsIII seemed to be somewhat more toxic than AsV, this effect was not significant. Boro and Aman cultivars also did not give a significantly different response in terms of shoot height. A strong relationship between RTI and RSH was found. Dasgupta et al. (2004) reported, at 1 mg/L AsV, a root elongation inhibition of 90% for rice cultivar Azucena and 50% inhibition for Bala.

Carbonell-Barrachina et al. (1997) found that AsIII was more toxic to bean plants than to tomato plants. Exposure resulted in a reduced dry biomass production with fruit dry biomass as the most sensitive parameter. The lowest AsIII concentration that caused significant toxicity was 2 mg/L for bean plants and 5 mg/L for tomato plants. Carbonell-Barrachina et al. (1999) exposed turnip to 0, 1, 2, and 5 mg/L MMA, DMA, AsV, or AsIII. MMA was the most toxic form, causing ~70% reduction in root dry matter production at 1 mg/L. Other As species were not toxic at this level. For AsV and AsIII, no toxicity was observed at the highest exposure level.

In two similar studies, marsh grasses (*Spartina patens* and/or *S alterniflora*) were exposed to various levels of DMA, MMA, AsIII, and AsV (0, 0.2, 0.8, 2.0 mg/L) (Carbonell et al. 1998; Carbonell-Barrachina et al. 1998). MMA was the only species toxic at 0.2 mg/L. The general conclusion was that MMA caused toxicity at the lowest tested concentration of 0.2 mg/L, whereas AsV only caused toxicity at 2.0 mg/L. In the presence of 0.2 mg/L inorganic As, dry matter production was increased. The As species were ranked according to their toxicity as follows: MMA, DMA > AsIII > AsV. The authors concluded that the form of As was more important than the exposure concentration.

High intraspecies variation for As tolerance was found for common velvetgrass (*Holcus lanatus*) (Hartley-Whitaker et al. 2001). EC_{50} values for root growth inhibition varied from 0.225 mg/L for the most sensitive clone to >75 mg/L for the most tolerant clone.

Bleeker et al. (2003) tested AsV tolerance of two populations of striated broom (*Cyticus striatus*), one adapted to an As-rich environment and the other one not. Plants were exposed for 7 d to 0.6, 1.2, 2.4, and 4.8 mg/L at two PO_4^{3-} levels. The lowest concentration causing root growth inhibition was 0.6 mg/L for the nonadapted population and 1.2 mg/L for the adapted population. The presence of PO_4^{3-} increased As tolerance, particularly for the nonadapted population. Sneller et al. (1999b) did short-term (3 d exposure) toxicity experiments with maiden's tears (*Silene vulgaris*) that revealed severe root growth inhibition (~75%) at the lowest test concentration of 0.188 mg/L in the presence of a low level of PO_4^{3-}. At high

PO_4^{3-}, the no observed effect concentration (NOEC) was 0.578 mg/L and the EC_{50} was 2.18 mg/L.

Spiked Soil. Onken and Hossner (1995) applied an AsIII or AsV addition of 25 mg/kg to the soil before the experiments. No time was given to reach steady state, which may explain the strong fluctuation in pore-water concentrations with a downward tendency. In the soils without As application, the As concentration in the soil solution (only AsV was found) gradually increased in time and reached equilibrium within 10 d after the increase started. In the silt loam soil, the increase started immediately after flooding, while the clayey soil had an lag time of 30 d before the increase started. In the silt loam soil, reduced dry matter was first observed after 40 d exposure. At the termination of the experiment (60 d exposure), the dry matter was reduced approximately 50%. There was no significant difference in response to AsV and AsIII. In the clayey soil, no toxicity was observed, suggesting that a greater part of the added As was not available. Taking into account the large uncertainties and fluctuations in soil-water concentrations, water from the clayey soil contained 10–15 times less As; plants grown in the silt loam soil contained 2–3 times more As.

Lambkin and Alloway (2003) grew barley (*Hordeum vulgare*) on soil spiked with AsV (0, 6, 40, 80, 160, or 320 mg/kg). The background concentration was 13 mg/kg. Germination was delayed at 80 mg/kg and higher. With increasing soil concentrations, the dry matter yield decreased. Most sensitive was seed yield. At 6 mg/kg, yield was reduced with 65%. In general, total dry matter plant yield was greatly reduced at 40 mg/kg and higher.

Jahiruddin et al. (2004) spiked a silt loam soil (background level, 2.6 mg/kg) with AsV: 0, 5, 10, 15, 20, 25, 30, 40, and 50 mg/kg. First a Boro rice cultivar (BRRI dhan 29) and then an Aman cultivar (BRRI dhan 33) were grown. For Boro, first significant effects occurred at 10 mg/kg soil, causing a grain yield reduction of more than 45% (Fig. 1). The As

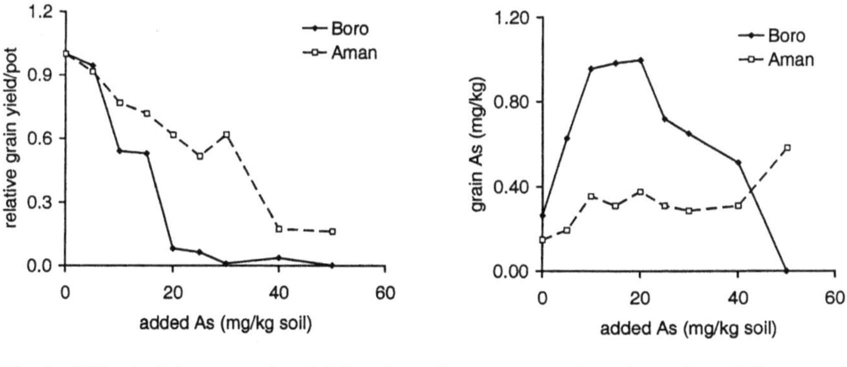

Fig. 1. Effect of As on grain yield and on As concentrations in grains of Boro and Aman rice cultivars grown consecutively in the same pots that were contaminated before experiments (Jahiruddin et al. 2004).

concentration in grains of Boro first increased with the exposure level but then decreased. Perhaps the toxic effects became so severe at that stage that As was hardly translocated anymore (Fig. 1). Other affected parameters were plant height, the number of panicles, and the straw yield. At 20 mg/kg soil, grain weight and the number of grains per panicle became significantly reduced, and the number of tillers was reduced at 30 mg/kg soil. For Aman, the first significant adverse effects were on the number of grains per panicle and straw yield at 10 mg/kg. At 20 mg/kg soil, grain yield became affected, whereas the other parameters were not significantly affected below 40 mg/kg soil. Overall, Boro suffered more than Aman, for which several explanations may be possible. (1) Aman was exposed to lower concentrations because of aging. (2) Part of the added As was removed from the system via biomass removal and/or volatilization. (3) The Aman cultivar was more tolerant by nature. In terms of QA/QC, the study had limitations. For example, no measures were described to remove possible external contamination of plants, which may explain the unlikely high Fe concentrations in grains of ~100 mg/kg whereas concentrations in rice are usually ~5 mg/kg. The chemical analysis included no certified reference material (CRM).

Soil Culture Irrigated with As-Contaminated Water. Abedin et al. (2002b) exposed rice cultivar BR11 to AsV and studied growth and As uptake (30-d-old seedlings; 170 d exposure; flooded soil conditions; soil from Scotland; 7 AsV levels: 0, 0.2, 0.5, 1.0, 2.0, 4.0, 8.0 mg/L; 2 PO_4^{3-} levels: 14.3 and 28.8 mg PO_4^{3-}/kg; greenhouse). The first observed adverse effect was reduced root biomass at 0.2 mg/L. Other effects including reduction of plant height, spiklet weight, number of spiklets, and grain yield started at 2 mg/L. At all exposure levels, no reduced yield of straw was observed.

In a similar experimental setup (BR11, greenhouse, full cycle, AsV at 0, 1, 2, 4, 8 mg/L, 14.3 and 28.6 mg/kg PO_4^{3-}, same soil), reduced root biomass, grain number, and grain weight (g/pot; 26% reduction) was found at ≥1 mg/L (Abedin et al. 2002a). With increasing AsV exposure concentration, other effects were also observed (≥2 mg/L: reduced straw weight, reduced 1,000 grain weight; ≥4 mg/L: reduced plant height; ≥8 mg/L: reduced tiller number). Comparing the two studies, irrigation water concentrations associated with toxic levels seem to deviated substantially despite the similar setup. The main reason is probably the difference in the lowest test concentration, which was 0.2 mg/L in Abedin et al. (2002b) and 1.0 mg/L in Abedin et al. (2002a). In both studies, first effects occurred at the lowest test concentration, indicating that the lowest test concentrations were too high to assess the NOEC and the lowest observed effect concentration (LOEC). It seems that for this particular experimental setup, the LOEC was equal or below 0.2 mg/L in irrigation water. Other differences such as the reduced straw weight found in Abedin et al. (2002a) but not in Abedin et al. (2002b) could not be explained.

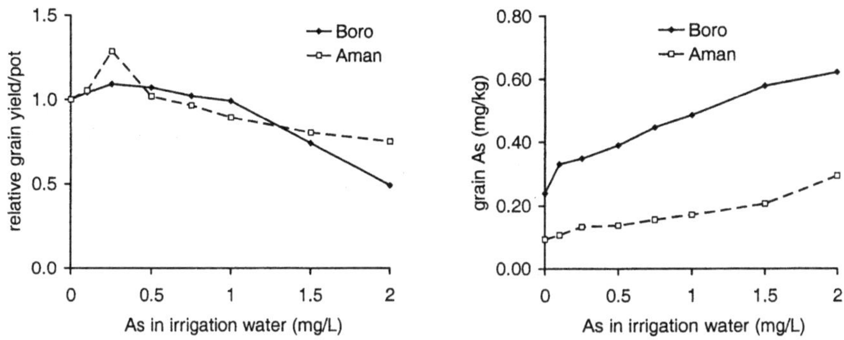

Fig. 2. Effect of As on grain yield and on As concentrations in grains of Boro and Aman rice cultivars grown consecutively in the same pots that were irrigated with contaminated water only during the Boro cultivation (Islam et al. 2004b).

Islam et al. (2004b) did a similar experiment with the same soil and rice cultivars as Jahiruddin et al. (2004) with the difference that AsV was now added via irrigation water (0, 0.1, 0.25, 0.5, 0.75, 1.0, 1.5, 2.0 mg/L) during Boro cultivation. During the Aman cultivation, As-free irrigation water was used. With an increase in As concentration in the irrigation water, first an increase in grain yield was observed, both for Boro and Aman. After that, yields declined (Fig. 2). As concentrations in grains steadily increased with As levels in irrigation water (Fig. 2). Within the tested range of As concentrations in irrigation water, the observed toxic effects and As accumulation in grains were far less compared to the observations within the range of soil concentrations used in (Jahiruddin et al. 2004). Similar patterns were observed, however, but it is not possible to determine in either experiment to what levels the plants were truly exposed.

In conclusion, none of the existing toxicity data can be regarded as representative for the field situation and extrapolations are not yet possible. A better understanding of As in the soil in relation to uptake and toxicity is therefore urgently needed. Ideally, soil parameters should be identified that correlate with uptake and toxicity. Emphasis must be given to the development of a methodology for toxicity experiments that can be extrapolated to the field.

Toxicity to Microorganisms. Soil microorganisms may also be affected by As toxicity (Mahimairaja et al. 2005). Effects of As and on the soil microbial community can be expected, with AsIII being more toxic than AsV. Microbes can adapt to As contamination, but this can be accompanied by a change in density and structure of the community. Ghosh et al. (2003) reported that microbial biomass and activity were negatively correlated with total and bioavailable As in soil samples from West Bengal. However, the description of the soils used was limited, making it difficult to assess if there were any other reasons such as different soil types and land use that could explain the results.

C. Foods

Inorganic As is generally recognized as the more toxic form to humans than organic As. Considering the presence of both forms in multiple foods, a well-balanced evaluation of As risk in foods should thus be based on inorganic As contributions and not on total As.

As Speciation in Foods. The methodology to assess As speciation in plant and animal tissue is complicated and still under development. A standardized methodology is still not yet available, and As speciation measurements depend on the pretreatment, extraction technique, storage of samples, and analytical method. Available values should therefore be regarded as experimentally defined levels of inorganic As species. The complicated methodology is also the main reason that CRMs for As speciation have not been developed except for some organic As species in marine products (CRM DORM-2: dogfish muscle). Table 1 provides an overview of the speciation data on rice, but the data presented should be carefully interpreted.

Most of the papers on As speciation in foods focus on method development. Pizarro et al. (2003b) tested three techniques to extract As species from rice. Two consecutive extractions with a 1:1 methanol:water extractant gave the best result with a recovery of >95% of total As in rice. Extraction with TFA was discarded because of the low quality of the chromatograms. Water only as an extractant was also discarded, although it gave more or less the same recovery as the methanol: water extractant. Adding methanol should improve the dissolution of organic As and avoid microbial activity during storage of extracts. As species in the 1:1 methanol:water extract of rice were stable for at least 2 mon. The authors showed that pretreatment and storage influence speciation (microbial activity, light induced, thermal degradation). Irradiation and grinding especially had a great influence. Irradiation avoids microbial activity, ensuring stability of the As species, but it may alter speciation. Pizarro et al. (2003b) showed that irradiation caused a decrease in DMA, AsV, and especially arsenobetaine (AsB), in favour of AsIII and MMA. The authors suggested that this may explain that the irradiated NIST 1568a does not contain AsB. As in flour, rice seemed to be more sensitive to microbial activity than As in whole grains during storage. Within 1 mon storage of flour rice, especially MMA and AsV decreased whereas AsIII increased. This observation will complicate the development of a CRM for As species, because grinding is necessary to certify homogenization while antimicrobial measures such as irradiation have to be taken to ensure stability of As species. Some discrepancies in the presented data were observed. The total As concentration

Table 1. As speciation data for rice.

Country	Type of rice	Total As µg/kg	Inorganic As (µg/kg)	Extraction efficiency (%)	Inorganic As (%)	Reference[a]
Spain	Rice	—	0.062	95	—	1
Spain	Paella rice	0.17	0.08	78	48	9
Italy	Rice	—	0.061–0.069	95	—	1
Italy	Various	0.19–0.22	0.10–0.14	77–103	53–65	9
USA	White rice	—	<0.025–0.271	—	—	2
USA	Rice	0.303	0.074	—	24	3
USA	White rice	0.21–0.34	0.021–0.095	86–97	11–35	4
USA	Brown rice	0.160	0.098	99	62	4
USA	Various	0.11–0.40	0.05–0.14	59–90	20–59	9
India	Various	0.03–0.08	0.02–0.05	62–88	36–67	9
Bangladesh	Various (Aman)	0.03–0.30	0.01–0.21	51–98	34–86	9
Thailand	Various	0.11–0.20	0.06–0.10	72–97	44–74	9
Unknown	Rice	0.410	0.367	94	90	5
CRM	NIST 1568a	—	0.087	—	31	6
CRM	NIST 1568a	0.280	0.092	94	34	4
CRM	NIST 1568a	0.283	0.085	99	35	7
CRM	NIST 1568a	0.290	0.080	83	33	9
CRM	NIST 1568a	0.286	0.088	99	31	10

—, No data.
CRM, Certified reference material for total As in rice.
[a]1, Pizarro et al. (2003b); 2, Lamont (2003); 3, Schoof et al. (1999); 4, Heitkemper et al. (2001); 5, Kohlmeyer et al. (2003); 6, Pizarro et al. (2003a); 7, Pizarro et al. (2003b); 8, Cava-Montesinos et al. (2003); 9, Williams et al. (2005); 10, Sanz et al. (2005).

extracted with methanol:water is not in agreement with the sum of concentrations of the individual As species. Therefore, the claimed 95% recovery with methanol:water is not fully justified; this also makes it impossible to calculate the percentage contribution of each As species to the total As concentration extracted with methanol:water. If the presented speciation data are correct, the analyzed rice contained 60–70 µg/kg of inorganic As. Total As ranged between 0.14 and 0.28 mg/kg raw weight. The moisture content varied (3%–15%), but so far as can be deducted from the paper no corrections have been made. All nonirradiated samples contained some AsB, which has not been previously reported.

The same authors tested water, methanol: water, and phosphoric acid as extractants for rice, fish, chicken (Pizarro et al. 2003a). For rice, fish, and chicken, the 1:1 methanol: water extractant (three consecutive extractions) was the best in terms of recovery and sample handling. For rice and fish, the recovery was ~100%, for chicken 75%. Phosphoric acid (1 M) was the best extractant for soil with a recovery of ~100%. Ranked according to abundance, rice contained AsIII > DMA > AsV > MMA. This ranking is in agreement with the results for NIST 1568a. No quantitative data on As species concentrations in rice was provided. Chicken contained DMA, AsB, and an unknown As species. Fish contained AsB and an unknown As species. Of the two soils tested, both contained mainly AsV (~80%) and some AsIII. One of the soils also contained some DMA and MMA. With respect to stability during extraction, there were some minor differences in extraction efficiency between the extractants, but the ratio between the species was more or less the same. This result was found for all matrices and indicates that speciation did not alter during extraction, which was confirmed by the observation that the recovery of standard additions of all identified As species to raw samples before extraction was always more than 90%. Stability tests during storage of extracts showed that after 3 mon storage (1:1 methanol: water for rice, fish, and chicken and 1 M phosphoric acid for soil), the As species in rice extracts were still stable. However, part of the AsB in fish and chicken was transformed to DMA whereas in soil part of the AsIII was oxidized to AsV. For fish, adding more methanol to the mixture may keep the extract stable up to 2 mon. For chicken, the extracts were more or less stable up to 2 mon.

Kohlmeyer et al. (2003) analyzed a great number of rice and seafood samples. For rice, an enzymatic digestion with α-amylase followed by drying and dissolution in water was used. Standard additions before extraction gave good results, and the results for NIST 1568a were in agreement with other reports. Rice was pulverized before extraction and analysis, but the storage time and its effects on speciation were not specified. In general, three As species were found in rice: AsIII, AsV, and DMA. The percentage of inorganic As was usually at least 50%, with maximum values of more than 90%. Of the 180 rice samples analyzed (origin unknown), total concentrations were between 0.05 and 0.5 mg/kg fresh weight. A typical raw rice sample contained 0.170 mg/kg AsIII, 0.193 mg/kg AsV, and 0.023 mg/kg DMA, whereas MMA was below detection limit. A typical parboiled rice sample contained 0.102 mg/kg AsIII, 0.010 mg/kg AsV, and 0.044 mg/kg DMA, whereas MMA was below detection limit. No AsB was found. Raw rice and brown rice had a higher total As and a higher percentage of inorganic As, mainly because of a higher AsV level, compared to white and parboiled rice. According to the authors, this may suggest that parboiling and/or polishing removes As from the rice and that the As is mainly present in the outer husk and bran layer. For seafood, a methanol : water extraction was used (once with 3 : 1, followed by two extractions with 1 : 1). Recovery of AsB and tetramethylarsonium was within the certified value of CRM DORM-2 (dogfish muscle). Marine fish mainly contained AsB (90–100%) and no inorganic As was found. Arsenosugars were predominant in marine algae. High concentrations of AsV were found in some brown algae such as *Hizikia* (25.6 mg/kg dw, which is 60% of the extractable As). High AsV was also found in a sample of roasted seaweed (12 mg/kg dw, which is 86% of the extractable As). The extraction efficiency of marine algae was highly variable; e.g., ~70% recovery for *Hizikia* and ~35% for roasted seaweed.

Lamont (2003) analyzed inorganic As in white rice (40 samples) from the U.S. with the aim to statistically describe the distribution of inorganic As concentrations in rice. Specialty rice, unpolished rice, precooked and parboiled rice, glutinous rice, "rice" that does not belong to the *Oryza* genus, and rice mixes were excluded. The rice samples taken into account were all milled to remove the hull, bran layers, and germ and were grown in the U.S. Inorganic As was extracted with HCl and brought via various steps into ammonium hydroxide, which was then acidified with formic acid and ready for analysis. Inorganic As concentrations ranged from <0.025 to 0.271 mg/kg ww (raw, uncooked), and the geometric mean was 0.112 mg/kg inorganic As. Based on statistical analysis of the data distribution, the author suggested that inorganic As concentrations in rice exceeding 0.250 mg/kg ww would represent extraordinary agricultural conditions or events. A drawback of the study is in the lack of detail in the methodology and QA/QC procedures. Also, total As concentrations were not measured.

Heitkemper et al. (2001) presented a method to determine As species in rice. Recovery of trifluroacetic acid (TFA) extractable As was 84%–99% of the total As. Total As concentrations were between 0.11 and 0.34 mg/kg, and inorganic As ranged from 0.021 to 0.104 mg/kg ww. The percentage of inorganic As was 11%–91%, and the remaining As was mainly DMA. Various other methods were tested (water extraction, methanol : water extraction, enzymatic extraction), but TFA provided the best results. Extraction with α-amylase followed by 1 : 1 methanol : water gave a 97% recovery of NIST 1568a, but the recovery of As from a rice sample was only 59%. The authors recognized that this method may become suitable with some more development, but considering the relatively long time enzymatic digestion requires and the good results obtained with TFA, the enzymatic extraction method was discarded. However, an advantage of the latter is that both AsIII and AsV can be quantified, which is not possible with TFA as part of the AsV is reduced to AsIII. Before extraction, rice was ground and stored at –10°C. No information was provided on storage time and stability of As species during storage. For all tested methods, the recovery of NIST 1568a was always better than for the samples, which could not be explained. Data from Pizarro et al. (2003b) showed the same trend. The influence of storage after extraction was tested, and although they were relatively stable,

Heitkemper et al. (2001) recommended that the extracts be analyzed soon as possible after preparation. The IC separation procedure did not separate AsIII and AsB, which was acceptable according to the authors because they did not find any report on AsB in rice. However, Pizarro et al. (2003b) did report AsB in rice. In spite of that, the separation is acceptable because an alternative IC procedure that did separate AsB and AsIII gave comparable results.

Schoof et al. (1999) extracted As species from 40 food commodities with 2 N HCl at 80°C for 16 hr. According to the authors, recovery of the four tested components (AsIII, AsV, MMA, DMA) was good and all components were stable throughout the procedure, which was shown by using standard additions. Raw rice was analyzed whereas other foods were prepared, i.e., ready for consumption. Rice contained the highest amount of inorganic As, 0.074 mg/kg ww ($n = 4$) whereas the total As was 0.303 mg/kg ww. All other commodities such as fish, meat, dairy products, vegetables, and fruit contained less than 0.011 mg/kg ww. The authors concluded that the concentrations are likely to be representative for the U.S. However, this is disputable because the data were not only derived from one sampling event and one state, which was acknowledged by the authors, but also because only four samples of each commodity were analyzed. There were some discrepancies in the data: the sum of As species was generally far below the total amount of As. The authors mentioned that the recovery of standard additions of As species was 86%–98% and that the recovery of total As from CRM DORM-1 was in agreement with the certified value. Therefore, if the extraction from the foods was indeed complete, there must have been one or more unidentified As species in the samples.

For seafood and marine plants, Li et al. (2003) reported that total As concentrations were 1.7–19.3 mg/kg dw in red algae, 14.6–38.7 mg/kg dw in brown algae, and 0.086–7.54 mg/kg ww in marine fish and shellfish. Fish and shellfish contained less than 2% inorganic As, whereas inorganic As was not detected in either brown or red marine algae. The latter is in contrast with Kohlmeyer et al. (2003), who found high inorganic As levels in some algae. Fish mainly contained AsB whereas As sugars were predominant in algae. As was extracted with 10 mL 1:1 methanol:water, repeated five times. The combined extracts were dried and dissolved in water and stored at 4°C. It is unclear whether the extraction method and storage were tested for As species stability. Extraction efficiency was mostly between 80% and 90%. Li et al. (2003) mentioned that although arsenosugars are not cytotoxic or mutagenic, arsenosugars in algae are metabolized after ingestion to other As species including DMA, which is more toxic than As sugars.

Huang et al. (2003) studied As speciation in farmed fish (*Oreochromis mossambicus*) in As-affected areas in Taiwan. The fish prefers brackish waters and are cultivated in ponds. Although contaminated groundwater is no longer used for drinking water, it is still a source of water for aquaculture. A 16 hr methanol:water (1:1) Soxhlet extraction was used and the extraction efficiency was ~85%. Results showed that there was a positive correlation between As in the water and fish. The water mainly contained AsV, whereas AsB predominated in the fish. Total As concentrations in fish ranged from 18 to 329 mg/kg dw, and inorganic As was ~5 mg/kg dw (range, 1.7–26.1 mg/kg dw).

Williams et al. (2005) presented speciation data for rice from various countries including the first data for India and Bangladesh. The results are discussed in more detail in section III. B. TFA was used to extract As species and gave a reasonably good recovery (83% ± 13%) for NIST 1568a. Although NIST 1568a is not certified for As species, the results are in good agreement with others (see Table 1). It seemed that the variability of the extraction efficiency was lower within rice varieties than between cultivars.

Recently, Sanz et al. (2005) presented a novel and rapid method for As speciation in rice. An enzymatic treatment (protease XIV and α-amylase) in combination with an ultrasonic probe resulted in a 3-min procedure giving a 90% extraction efficiency. Spiking experiments showed that As species remained stable throughout the procedures. Results for CRM NIST 1568a were in agreement with those from others.

III. Situation in Bangladesh

A. Agriculture

Irrigation Water. During the last three decades, much land has been brought under Boro rice cultivation in the dry season by using irrigation shallow tube wells (STWi), which is one of the main reasons that the country is self-sufficient in rice production today. Boro rice receives the most irrigation water of all crops, with an estimated amount of 1,000 mm/cycle. The total area under irrigation is 4 million ha, and 75% is covered by groundwater resources: 2.4 million ha via 924,000 STWi and 0.6 million ha via 23,000 deep tube wells for irrigation (DTWi). In the dry season, 3.5 million ha is used for Boro, 0.23 million ha for wheat, and 0.27 million ha for other crops. Classifying the divisions according to the area under irrigation gives the following ranking: Rajshahi (39%) > Dhaka (27%) > Chittagong (13%), Khulna (12%) > Sylhet (7%) Barisal (2%). The area under Boro rice production follows the same pattern. Wheat and other crops follows a somewhat different pattern, with Rajshahi as the most important area followed by Dhaka and Khulna (BADC 2004).

With regard to contaminated STW used for drinking water (STWd) and arsenicosis, the As-affected areas are mainly located in the south/southwest, i.e., Khulna, Dhaka, and north Chittagong. With an estimated 25% of the STWd having As concentrations above 0.050 mg/L, it can be expected that a high percentage of STWi also has high levels of As. The percentage is unknown, because the spatial distribution of irrigation wells deviates from drinking water wells. In groundwater, As is mainly present as AsIII and AsV, and the amounts are in the same order of magnitude.

Data from Jessore (JICA/AAN 2004) showed that 87% (74 of 85 tested wells) of DTWi contained more than 50 μg/L. Average As concentration of those wells was DTWi was 0.21 mg/L) This value is very high because DTWs are generally considered to be safe for consumption (<0.050 mg/L). One of the probable reasons for this is that those so-called DTWi are actually shallower than the DTWd in the same area (~100 versus 200 m). Of the STWi that were at the same depth as the STWd, 24% (59 of 246 tested wells) contained more than 0.050 mg/L. The average concentration of those wells was 0.07 mg/L.

The terminology STW and DTW has been confounded in the past. The distinction between DTW and STW was based on the capacity of the pump: STW pumps use suction-mode (centrifugal) pumps, have a maximum lift of about 7–8 m, with pipes used that are less than 10 cm in diameter and irrigate about 4 ha. DTW pumps can in principle tap groundwater from any depth, use force-mode pumps, have pipes up to 25 cm in diameter, and irrigate up to 25 ha.

Farid et al. (2005) studied seasonal and temporal variation of As concentrations in a single STW. During the monitored Boro season, a small seasonal effect was observed. Starting in January, the concentrations slowly increased, reaching the highest concentration in early March, and then declining again, reaching the original level in June. The difference between the highest and lowest concentration was, however, only 5%. Diurnal variation was not found.

It has been estimated that 900,000–1,360,000 kg As/yr is brought onto the arable land via groundwater extraction for irrigation (Ali 2003). Especially in Southwest and Southcentral Bangladesh, the deposition of As on the arable land is high. Northwest Bangladesh, which has relatively low As concentrations in the shallow aquifer but has a very high intensity of STWi, is also extracting a considerable amount of As from the aquifer. Other sources such as P fertilizer and manure are likely to be minor sources of As, but this needs confirmation. Based on existing national data for As in STW used for drinking water and the distribution of Boro rice production, Duxbury and Zavala (2005) estimated that 10 yr of irrigation with As-contaminated water would add 5–10 mg/kg soil to 41% of their 456 study sites (all paddy fields) in Bangladesh. In a case study in West Bengal, India, data on As in irrigation water and the paddy soil profile indicated a yearly As input of 1.1 mg/kg to the topsoil (Norra et al. 2005). According to Meharg and Rahman (2003), 150–200 (up to 900) mm water is used for land preparation before rice planting and crop growth requires 500–3,000 mm. As a conservative estimate, they assumed 1,000 mm groundwater/yr (1,000 L/m^2/yr). If the irrigation water contained 0.1 mg/L and the As was retained in the first 10 cm of soil (assuming a soil density of 1 kg/L), the water input would result in a yearly increase of 1 mg As/kg soil.

Ross et al. (2005) estimated that 76% of the Boro is grown in areas where STW usually contain less than 50 μg/L, 17% in areas with 50–100 μg/L, and 7% in areas with more than 100 μg/L. These estimates suggest that the bulk production of Boro, which is largely distributed to Dhaka, takes place mainly in areas with low As in the shallow aquifer. Although less Boro production takes place in the areas with high As in the shallow aquifer, the rice that is produced is likely to be used for personal/local consumption.

Recently, the Bangladesh Agriculture Development Corporation has started to monitor As in irrigation water wells on a national scale. Their activities also plan to include monitoring As levels in crops and soils.

Soil and Crops. Meharg and Rahman (2003) conducted a preliminary survey of As in rice and soil from Bangladesh. A total of 71 soil samples were collected throughout the country. The highest measured soil concentration was 46 mg/kg, whereas less than 10 mg/kg was found in areas with low As in irrigation water. West Bangladesh seems to have the highest soil concentrations (>30 mg/kg), followed by the central belt, which is in agreement with groundwater concentrations. At various locations with high As levels in groundwater, low concentrations were found in the soil. However, this was not always the case. A correlation between soil concentrations and irrigation water concentrations was observed when the age of the water well was taken into account. The data also indicated a positive relationship between As concentrations in rice and soil.

In agreement with Meharg and Rahman (2003), data from a preliminary nationwide survey of As in soil, crops, and irrigation water indicate that the west-southwest part of Bangladesh contains the highest As concentrations (Miah et al. 2005). In these parts, irrigated soils had higher levels of As compared to adjacent nonirrigated soils. In the irrigated soils, the first 0–15 cm had the highest levels of As. In other parts of the country, irrigated and nonirrigated soils did not differ in As concentrations. The differences in soil concentrations were however not reflected by As levels in the rice plants.

Islam et al. (2005) studied As levels in water, soil and crops at 456 locations in five subdistricts. The average As concentration in the soil was 12.3 mg/kg (range, 0.3–49 mg/kg), and the subdistricts were classified according to soil concentrations: Faridpur > Tala > Brahmanbaria > Paba > Senbag. Of all soil samples, 53% contained <10 mg/kg, 26% contained 10.1–20 mg/kg, and 18% contained >20%. Concentrations were highly variable both between and within subdistricts. The same was observed at the command area and paddy field level. In some cases, this correlated with the distance to the tube well, in other cases the variation seemed to be random or related to microelevation. Islam et al. also found a high seasonal variation in As soil concentrations: at the end of the Boro season the soil concentration had increased sharply when irrigated with As-rich water, but that most of it was again removed after the Aman (flooding) season. There are various possible explanations for this observation. (1) During the rainy season, a substantial area of Bangladesh is flooded with rainwater for a couple of months, at some locations up to 3 m depth. Part of the standing water leaches downward, transporting As to deeper layers. (2) As desorbs to the standing water and is then removed laterally. (3) the top layer may be eroded and run off during heavy rainfall. (4) Volatilization of As may occur during periods of flooding. The general opinion is that leaching is an unlikely process because of the slow percolation rate, 2–4 cm/d (Brammer 2005; Islam et al. 2005). Experiments done with undisturbed soil columns have confirmed this (Khan 2005), explaining that soil concentrations in the first 15 cm are generally highest compared to the rest of the soil column.

Islam et al. (2000) reported, for some soil samples from Nawabgonj, Rajarampur, Jessore, Jhenidah, and Comilla, total As concentrations of 5–33 mg/kg with an average of 17 mg/kg. However, the study has some limitations. No information was provided about the use of the sampling locations (e.g., fallow land, paddy field, other crops, residential area, etc.). Therefore, the relation between As in groundwater and the topsoil cannot be discussed. Further, no explanation was given for the low extraction efficiency (<50%) of the acid digestion compared to the total analysis with X-ray fluorescence spectroscopy (XRF). The use of CRM was not mentioned in the paper.

Jahiruddin et al. (2005) collected 100 samples of irrigation water, soil (composite sample of five randomly taken samples at 0–15 cm depth), and rice plants at 100 STW command areas in Chapai Nawabganj (Sadar subdistrict) during the Boro season of 2003. The irrigation water contained 0.025–0.352 mg/L (mean, 0.075 mg/L). Soil concentrations were 5.8–17.7 mg/kg (mean, 11.2 mg/kg), straw contained 1.48–17.6 mg/kg (mean, 5.88 mg/kg), and rice grain contained 0.241–1.298 (mean, 0.759). Poor correlations were observed between grain and soil and between grain and water; a good correlation was found between grain and straw. The total soil As was correlated with As in irrigation water. Total As was not correlated with any soil parameter. In terms of quality, the study had some limitations. Rice cultivars were not identified, pretreatment of straw and grains (e.g., rinsing with distilled water to remove dust) was not mentioned, and CRMs were not included.

Five soil profiles of 15 m depth were collected at Deuli village (near Samta village, Southwest Bangladesh) (Yamazaki et al. 2003). Although not specifically mentioned in the paper, it seems the samples were collected from locations that are not being used for agriculture. Results showed that the soil concentrations were dependent on the type of sediment. Sandy sediments contained 3–7 mg/kg (median, 5 mg/kg) and clay sediments contained 4–18 mg/kg (median, 9 mg/kg), whereas peat and peat-clay sediments contained 20–111 mg/kg. The first 6 m contained sandy and clay layers while the peat layers were found at a depth of 7–10 m. Below 10 m, a sandy layer was dominant. The general pattern was that the As concentrations were relatively constant except for the 7–10 m layer, which contained much higher total As. The profile of the 1 M HCl extractable As gave more or less the same pattern although the topsoil also had a relatively high amount of extractable As. The total As and the extractable As were well correlated.

Twenty-five locations representing five subdistricts (Chapai Nawabganj Sadar, Kustia Sadar, Bera, Ishurdi, and Saishabari) of four districts were sampled at three depths: 0–15 cm, 15–30 cm, and 30–45 cm (Alam and Sattar 2000). Soil concentrations ranged from below detection limit (not specified; lowest reported concentration was 1.3 mg/kg) to 56.7 mg/kg. Ten of the 25 locations contained As concentrations >20 mg/kg in at least one of the sampled layers. Of the total number of 75 samples, 18 samples contained more than 20 mg/kg, whereas 25 samples contained no As above the detection limit. As concentrations in the adjacent water wells ranged from below detection limit (not specified; lowest reported value was 0.010 mg/L) to 0.071 mg/L, i.e., not very high. A positive correlation was found between As concentrations in the soil and water. The authors also tried to correlate soil parameters (sand, silt, clay, pH, and OM) to the As concentration in the soil, but the results were inconsistent. In terms of quality, the paper had serious shortcomings. Land use of the sampling locations was not described, which is an important feature because irrigation depends greatly on the type of crop. It was not mentioned if the sampled water wells were drinking water or irrigation wells. In the methodology section, QA/QC was not described and CRM was not used. Techniques to determine soil parameters were not described, and data were not presented.

Breit et al. (2005) found, at a number of nonpaddy field locations, subhorizontal layers at less than 3 m depth enriched in Fe (up to 13%), Mn (up to 1.7%), and As (up to 760 mg/kg). These layers marked the upper limit of water-saturated sediment. Precipitation of Fe, As, and Mn was explained by oxidation at the border between nonwater-saturated conditions (brown sediments; oxidizing conditions) and water-saturated conditions (grey sediments; reducing conditions). Fe-enriched layers found under paddy fields were not enriched in As and Mn; however, this has not been explained yet.

Das et al. (2004) collected soil samples ($n = 18$) in three subdistricts: Kachua, Hajiganj (both in Chandpur district), and Sharishabari (in Jamalpur district). Composite soil samples (15–45 cm depth) were probably taken from arable land but specific land use was not mentioned. A CRM was not included in the soil analysis. The soil concentrations ranged from 7.3 to 27.3 mg/kg with an average of 15.7 ± 6.6 mg/kg. A positive correlation was found between the As in shallow tube wells and soil.

In neighbouring West Bengal (India), Domkal block, fallow lands contained 5.31 mg/kg while adjacent lands irrigated with 0.082 mg/L and 0.17 mg/L contained 11.5 mg/L and 28.0 mg/L, respectively (Roychowdhury et al. 2002a). The calculated input of As was approximately 1.6–16.8 kg/ha/yr. Good correlations were found between concentrations in STW and soils. On a paddy field scale, a good correlation between As in the soil and the distance to the STW was also found. The study clearly showed the effect of the contaminated irrigation water on As concentrations in the soil. Similar correlations were reported by Hossain (2005) for a number of locations in Faridpur, Bangladesh. In the latter study, they also found a relationship between micro-elevation and As soil concentration.

A study from West Bengal (India) showed further strong evidence for As buildup in topsoil because of irrigation with contaminated groundwater (Norra et al. 2005). Water, soil, and plant samples were collected from three fields: one paddy field and one adjacent wheat field, both irrigated with water containing 0.5–0.8 mg/L, and one reference paddy field not contaminated with As. Soil profiles were collected down to 110 cm depth. The upper topsoil in the contaminated paddy field contained 38 mg/kg, the less intensively irrigated wheat field grown with wheat contained 18 mg/kg, and a reference paddy field contained 7 mg/kg. The soil profiles of the contaminated paddy field and wheat field clearly showed decreasing As level with increasing depth. Although to a lesser extent, As did not only build up in the paddy field but also in the wheat field, which has not been reported before. Roots and shoots of two plants collected from the contaminated paddy field contained more As than the single plant collected from the reference site, but the grains did not. Neither the Boro rice variety nor the wheat variety was determined. According to the authors, continuation of irrigation with As-rich water could result in alarmingly high soil concentrations within a few decades. This conclusion has also been stated by Duxbury et al. (2003).

CIMMYT-Bangladesh, in collaboration with a number of national and international partners including the Bangladesh Agriculture Research Institute, Bangladesh Rice Research Institute, Bangladesh Agricultural University, Cornell University, and Texas A&M University, have done significant research on As in agriculture, focussing on the relationship of As in water, soil and crops. The final reports are not available at this writing.

Little work has been done on the potential risk of As in irrigation water to crop production. Appropriate field studies of possible effects on plant growth have not been carried out. With the high spatial variability in As concentrations in the soil, detailed studies at the microlevel are needed to investigate possible correlations between As in the soil and plant parameters. Duxbury et al. (2003) studied As concentrations in rice yields and panicle sterility in Bangladesh. They did not find any indication of toxic effects under current field conditions. Possible effects may however have been overlooked because of a lack of in-depth study.

To summarize, despite the limitations on quantity and quality of available data, there is clear evidence that soil concentrations are increasing over time because of irrigation with As-contaminated water. It still needs to be elucidated under which conditions, time frame, and scale soil accumulation takes place. In agreement with the recommendations from experts of the International Symposium on Behaviour of Arsenic in Aquifers, Soils and Plants held on January 2005 in Dhaka, long-term monitoring of As in soils under the various conditions present in Bangladesh in combination with detailed studies on the behaviour of As in paddy fields are required. The risk of long-term use of As-rich irrigation water to crop production has received little attention. A major area for future work is to generate reliable phytotoxicity data to evaluate current and future soil concentrations, particularly for flooded soil conditions.

B. Foods

To assess dietary exposure to As, data on As in foods, food consumption, and water intake are needed. It is advised against using total As concentrations for risk assessment purposes because these will give an overestimation of the risks. Because data on inorganic As in foods are very limited, some data on total As are presented here.

It can be expected that As concentrations in crops depend on the plant species, variety, and growth conditions. Therefore, data to assess the dietary exposure to As should be derived from the area/region for which the assessment is made. In other words, data from West Bengal may be applicable to Bangladesh, but data from the U.S. are unlikely to be representative for Bangladesh.

Inorganic As in Foods. Recently, the first data on As speciation in rice from Bangladesh have been published (Williams et al. 2005). The study examined As speciation in Aman rice from different countries. In Bangladesh, 15 different rice varieties were bought at the wholesale market in Dhaka and analyzed for total As, DMA, and inorganic As (AsV + AsIII). TFA was used for extraction, and the recovery ranged from 51% to 98%. The average total concentration was 0.13 ± 0.02 mg/kg (range, 0.03–0.30 mg/kg). Of the total As recovered with TFA, the average percentage of inorganic As was 80% ± 3% (range, 51%–98%). The level of inorganic As in rice from Bangladesh ranged from 0.01 to 0.21 mg/kg. For total As, concentrations were ranked as follows: U.S. > Europe > Bangladesh > India. Rice from Bangladesh and India had the highest percentage of inorganic As, resulting in more or less equal amounts of inorganic As in rice from the U.S. and Bangladesh. This indicates that the percentage of inorganic As in rice is not a constant factor and probably depends on cultivar and growth conditions. As speciation analysis on a number of vegetables from Bangladesh indicated that almost all As was present in the inorganic form (Islam and Meharg, unpublished data, 2005).

Total As in Foods. In Table 2, some data on total As concentrations in rice from Bangladesh and other countries are summarized. Highest concentrations were reported from districts with high As in soil and groundwater (Meharg and Rahman 2003). In this study, QA/QC was well described and included blanks, spiked samples, duplicate samples, and CRMs for soil and plants. Recoveries were 88% and 78%, respectively, for which data were not corrected.

Table 2. Total As concentrations (mg/kg dw) in rice from Bangladesh and West Bengal, India.

Country	Location	Rice (mg/kg)	Soil (mg/kg)	Remarks	Reference[a]
Bangladesh	Gazipur	0.092 (0.043–0.206)	10.9; 14.6	11 cultivars[b]	1
	Bogra	0.058–0.104	4.9–15.5	4 cultivars	1
	Dinajapur	0.203	11.7	BR11	1
	Naogaon	1.835	24.3; 26.7	BR11	1
	Nawabgonj	1.747; 1.775	15.7; 20.9	BR11	1
	Mymensingh	0.078	6.0–25.4	BR8	1
	Rangpur	0.185	6.5–11.5	BR11	1
	Rajshahi	0.075–0.117	7.8	3 cultivars	1
	Various	0.183 (0.108–0.331)		Raw Boro rice ($n = 78$); 14% water	2
	Various	0.117 (0.072–0.170)		Raw Aman rice ($n = 72$); 14% water	2
	Various	0.125		Processed rice ($n = 21$)	2
	Kachua, Hajiganj, Sharishabari	0.14 (0.04–0.27)	15.68 (7.31–27.28) (15–45 cm depth)	($n = 10$)	3
	Northwest	0.173		As-affected area	7
	Chapai Nawabganj	0.759 (0.241–1.298)	11.2 (5.8–17.7)	As-affected area; $n = 100$	8
	Market	0.13 (0.3–0.30)		Various Aman cultivars ($n = 15$)	9
West Bengal	Jalangi & Domkal	0.239 (0.043–0.662)		Raw rice ($n = 34$)	4
	Jalangi & Domkal	0.569 (0.198–1.930)		Cooked rice ($n = 18$)	4
West Bengal	South Parganas	0.072 ± 0.010		Precooked rice	5

[a]References: 1, Meharg and Rahman (2003); 2, Duxbury et al. (2003); 3, Das et al. (2004); 4, Roychowdhury et al. (2002b); 5, Mandal et al. (1998); 6, Kohlmeyer et al. (2003); 7, Watanabe et al. (2004); 8, Jahiruddin et al. (2005); 9, Williams et al. (2005).

[b]Lowest concentration in BR11: 0.043 mg/kg dw.

Roychowdhury et al. (2002b) found that cooked rice had approximately a twofold-higher level of As compared to raw rice, likely because of parboiling and/or boiling of rice in As-contaminated water. Bae et al. (2002) reported that As concentrations in rice after boiling in As-contaminated water (0.223–0.373 mg/L) were increased from 0.178 mg/kg dw to 0.228–0.377 mg/kg dw. On the other hand, Duxbury et al. (2003) found that processing (parboiling and milling) of rice reduced As concentrations ~20%.

Only three papers were found on total As in foods other than rice (Table 3) (Alam et al. 2003; Das et al. 2004; Roychowdhury et al. 2002b). They all described the methodology reasonably well, and CRM was

Table 3. Summary of As concentrations in various foods from Bangladesh and West Bengal, India (mg/kg dw).

Country	Item	n	Mean	Remarks	Reference[a]
West Bengal	Vegetables	77	0.123 bd, 0.690	Highest levels were found in leafy vegetables and potato skin.	1[b]
	Spices	25	0.208	0.013–0.481	1
	Miscellaneous	9	0.138	bd, 0.900	1
Bangladesh	Arum *Colocasia antiquorum*	9	0.09–3.99		2
	Potato *Solanum tuberisum*	5	0.07–1.36		2
	Water spinach *Ipomoea reptams*	6	0.68 0.1–1.53		2
	Various vegetables	32	<1.0	Bitter melon (*Momordica charantia*); gourd (*Trichosanthes dioeca*); okra (*Hibiscus esculentus*); data stem (*Amaranthus lividus*); Indian spinach (*Baseella rubra*); ribbed gourd (*Luffa acutangula*); cowpea (*Vigna sinensis*); jute leaves (*Corchoorus capsularis*)	2
	Ophicephal punctatus	9	<1.0		2
	Vegetables	30	0.05–0.5	Duplicate samples of 15 commonly grown and consumed types of vegetables sampled at Samta village	3

bd, below detection limit of 0.04 μg/kg.
[a]References: 1, Roychowdhury et al. (2002b); 2, Das et al. (2004); 3, Alam et al. (2003).
[b]To recalculate concentrations to wet weight, the authors used 5%, 80%, and 5% moisture content in rice, vegetables, and spices, respectively.

included. Most samples were collected from a few locations known for high As in the shallow aquifer, and the samples contained less than 0.5 mg/kg dw. Alam et al. (2003) mistakenly did not convert dry weight concentrations to wet weight before data interpretation, including estimation of the daily As intake (personal communication, E. Snow, 2004); this caused a fivefold overestimation of dietary As intake.

Das et al. (2004) reported that total As concentrations in fish were below 1 mg/kg dw. However, data from Taiwan showed that fish cultivated in As-rich water may have high levels of inorganic As (Huang et al. 2003); this will depend strongly on the fish species and size.

As speciation is likely to vary within and between plant species and with exposure conditions. Therefore, percentages of As species found in one study cannot be extrapolated to studies that only reported total As. This is in agreement with Williams et al. (2005), who found that rice from Bangladesh had a higher percentage of inorganic As than rice from the U.S. and Europe.

Food and Water Consumption in Bangladesh. Only one study was found reporting food consumption on a gram per capita per day basis in Bangladesh (Hels et al. 2003). That study reported data for two villages, Falshatia (Manikganj) and Jorbaria (Mymensingh), covering 1981–1982 and 1995–1996. Here, only the period 1995–1996 is summarized (Table 4), for 24-hr food weighing data collected in October–November and in January–March. Data were collected on a household level and consumption per capita was derived from that value; corrections were made for the number of meals consumed outside the house. All members of a household were treated equally. Seasonal variation in food consumption was observed, particularly for rice and vegetables. There were also differences between the villages, but the seasonal variation seemed to be greater. Both sampling events did not cover the fruit season, explaining the low fruit consumption.

Watanabe et al. (2004) studied water intake by adult men and women in two As-affected areas and reported a total water intake of 4.6 and 4.2 L/d respectively. Two methods were used to assess drinking water intake: (1) 24-hr self-reporting and (2) interview with frequent visits. The range of water intake was 1–6 L/d. The direct intake of drinking water was equal between men and women (~3 L/d). Water intake via food preparation determined by field experiments added another 1.6 L/d (men) and 1.0 L/d (women).

Mandal et al. (1998) estimated drinking water intake of a small number of people in an As-affected village; the average intake by adults was 4 L/d. The highest intake values for some individuals were 7 and 8 L/d. The methodology was not described. In contrast with Watanabe et al. (2004), a great difference in water consumption between sexes was reported. The general impression is that adults consume 3 L/d plus an additional 1 L/d for food preparation.

Table 4. Estimated food consumption for two villages in Manikganj and Mymensingh (g/capita/d) in 1995/1996.

Food group	Manikganj Oct–Nov $n = 152$[a]	Jan–Mar $n = 145$	Mymensingh Oct–Nov $n = 152$	Jan–Mar $n = 143$	Average
Rice	399	454	446	502	450
Nonrice cereals	20	20	2	12	14
Green leafy vegetables	28	49	46	37	40
Other vegetables	106	166	59	221	138
Roots and tubers	43	101	64	37	61
Fish	52	37	37	30	39
Animal products excluding fish	23	25	35	39	31
Fats and oils	7	5	10	9	8
Spices	5	6	8	6	6
Fruits	7	7	6	9	7
Miscellaneous	10	13	5	11	10
Total intake	701	887	719	914	804

From Hels et al. (2003).
[a]*n*, number of surveyed households; average household size, 5; average consumption unit, 4.5.

Published Human Exposure Assessments. To date, a few localized exposure assessments have been made in Bangladesh and West Bengal, all based on total As. Roychowdhury et al. (2002b) estimated the daily intake of total As via water and food for two locations in West Bengal (Table 5). Food consumption was subdivided into rice, vegetables, and spices, and averaged concentrations for these categories were used. Intake via foods was approximately 180 and 97 μg/d for adults and children (10 yr old), respectively. Via drinking water, adults and children were exposed to approximately 400 and 200 μg/d. No reference was given for the assumed drinking water consumption. Drinking water accounted for ~70% of the exposure whereas rice contributed ~30%. Compared to Hels et al. (2003), the food consumption data from Roychowdhury et al. (2002b) are high. Watanabe et al. (2004) estimated that the daily intake of total As by adults was approximately 600 μg/d (men, 674 μg/d; women, 515 μg/d), with 70% via drinking water and 10% via rice. The food consumption data seemed to be only rough estimates. In none of these studies was any variation in concentrations, consumption, seasonal effects, and As speciation taken into account (Roychowdhury et al. 2002b; Watanabe et al. 2004).

Duxbury et al. (2003) analyzed 150 rice samples from Bangladesh. Assuming a rice consumption of 400 g/d with 0.250 mg/kg and a water intake

Table 5. Published exposure assessments for Bangladesh/West Bengal. Consumption (g/d or L/d) with rice on dry weight basis and others on wet weight basis.

	Roychowdhury et al. (2002b)[a] Consumption[c]	[As]	Daily intake[c]	Watanabe et al. (2004)[b] Consumption[d]	[As]	Daily intake
Rice	750/750/400	~236	168/168/90	523/300	173	90/52
Vegetables	500/500/300	108	11/11/.5			
Wheat (bread)				179/131	362	63/46
Potato				130/66	5	0.65/0.33
Fish				72/26	830	60/22
Spices	10/5	~150	1.4/1.4/0.7			
Water	4/3/2[e]	116	466/350/233	4.6/4.0[e]	100	480/395
Total As intake			646/530/330			674/515
via food (μg/day)	180/180/97			214/120		
Water (%)			~70	~72		
Rice (%)			~30	~12		

[As], As concentration (μg/kg dw or μg/L).
[a]Average values for two As-affected areas in West Bengal.
[b]Results from two communities in an As-affected area in Bangladesh.
[c]Male/female/child (10/yr) (body weight 60/45/25 kg); rice, vegetables, and spices contained ~5%, 80%, and 5% water.
[d]Male/female (body weight ~55 kg); food consumption was based on a food frequency questionnaire survey.
[e]Including water for food preparation.

of 4 L/d with 0.050 mg/L As (drinking water standard, Bangladesh), the total daily intake would be 0.3 mg/d to which rice contributes 33% of the total; 14% of their rice samples contained ≥0.250 mg/d. Meharg and Rahman (2003) assumed a rice consumption of 420 g/d with 0.5 mg/kg and a water intake of 2 L/d (the WHO default value) with 0.1 mg/L. The calculated total daily intake was 0.41 mg/d and rice contributed 50% of the As. However, taking into account the climate in Bangladesh and the high consumption of rice, 2 L/d is likely to be an underestimation.

Guidelines and Standards. The hazards of As to human health are well known (WHO 2001). Dose–response relationships have been assessed, but these refer to As levels in drinking water and its health effects. The WHO guideline value is 0.010 mg/L whereas the Bangladesh drinking water standard is 0.050 mg/L. A provisional maximum tolerable daily intake of 0.002 mg/kg body weight/d (PMTDI; inorganic As only) was established in 1988, and this value is commonly used to evaluate dietary intake (WHO 1996). After 17 yr this value still has a provisional status. The Bangladesh drinking water standard and the PMTDI are not in agreement (Duxbury and Zavala 2005; Williams et al. 2005). Assuming a body weight of 60 kg, the PMTDI is 0.12 mg/d. A water consumption of 3 L/d with 0.050 mg/L would already exceed the PMTDI, regardless the levels of As in foods. This finding suggests that the PMTDI and the Bangladesh drinking water standard need to be reevaluated.

Many papers refer to the Australian food safety standard for total As of 1 mg/kg fresh weight to assess if As levels in sampled foods are acceptable (Abedin et al. 2002a,b; Das et al. 2004; Islam et al. 2004b; Jahiruddin et al. 2004; Norra et al. 2005). However, the validity of this comparison is questionable as that food safety standard has various limitations: (1) the Australian food safety standard is for total As only, ignoring the great differences in toxicity between organic and inorganic As species that commonly coexist in foods, and (2) in contrast with Australia, Bangladesh depends heavily on a single staple food, rice. An inorganic As concentration of 0.3 mg/kg in rice with a consumption of 0.4 kg rice/d would already equal the PMTDI, neglecting any other possible source of exposure. It is therefore advised against using the Australian food safety standard in the context of Bangladesh.

To summarize, based on the current available information a representative risk assessment for As in the food chain cannot be made due to lack in adequate exposure assessment. Data are needed on (1) inorganic As in foods from Bangladesh, (2) food consumption, including seasonal and spatial variability, and (3) water consumption, focusing on seasonal differences. Another point of concern is the inconsistency in guidelines and standards related to As exposure, making it difficult to evaluate the risks of As in foods. It cannot be excluded that concentrations in food crops will increase over time because of the prolonged input of As-contaminated irrigation water. With many efforts in progress to reduce exposure via drinking water, the relative importance of As in foods can be expected to increase.

C. Management Options

Risks to food safety and yield are likely to increase with the buildup of As in the soil. Although the risks cannot be quantified for the time being, it is proposed to focus on preventing and minimizing input via irrigation and uptake by crops (Brammer 2005; Duxbury et al. 2003; Lauren and Duxbury 2005; Meharg 2004; Panaullah et al. 2005; Ross et al. 2005). For example, many rice farmers in Bangladesh are using more irrigation water than needed. Optimizing water input could thus reduce input. Maintaining aerobic growth conditions in paddy fields, by using furrows, for example, may reduce bioavailability and uptake of As by rice. Breeding crops tolerant to As with a low accumulation of As in grains may reduce potential risks as well. Shifting from rice to less water demanding crops where possible is another alternative. Phytoremediation has also been suggested, but this is unlikely to be successful in an intensive agricultural system.

IV. Future Requirements

The use of As-contaminated irrigation water poses a risk to both food safety and food crop yield. Our knowledge is still limited in quantifying and managing the risks. To overcome this, a set of integrated activities needs to be carried out over a long-term period considering the chronic nature of As input to the agricultural system in Bangladesh. The outcome of such a program should provide direct input to all stakeholders, from policy makers to the farm level and the poor.

A. Agricultural Sustainability

In Bangladesh, agricultural production is under pressure from multiple factors, including arsenic. Recently, it has been noted that rice production per hectare is stagnating and at some locations declining. Deterioration of soil quality has been mentioned as one of the main factors, including nutrient depletion and loss of OM. Another increasing problem is access to water. Boro rice production depends heavily on irrigation with water from the shallow aquifer, but the water table is lowering rapidly at many locations because of excessive extraction. The high salinity of irrigation water is also hampering crop production in the coastal areas. The continuous addition of As to the topsoil may pose a long-term risk to soil quality and crop production and, therefore, to agricultural sustainability. The prevention of soil contamination deserves high priority as remediation of contaminated agricultural soils is extremely difficult. To assess and manage the risks of As to agricultural sustainability, the following activities are recommended as necessary.

- Establish a national water/soil/crop quality monitoring system to quantify the sustainability and long-term changes in the agricultural system, including not only As but also other parameters determining soil quality and crop production.
- Quantify the relationship between As in irrigation water, soil, and crops.
- Perform As mass-balance studies in the field to quantify input, accumulation, and output.
- Develop a phytotoxicity database for crops relevant to the field.
- Develop a predictive model for long-term changes in soil quality and crop production.
- Develop guideline values for As in irrigation water, soil, and crops.
- Develop a set of management options.

For policy makers, information such as the extent of As in irrigation water in posing long-term risk to agricultural sustainability and possible management options can be derived from these proposed activities. These data should be weighed against all other risk factors to set priorities, resulting in a clear and well balanced assessment of constraints and solutions.

B. Food Safety

A comprehensive human health risk assessment is needed in determining food safety levels. Although drinking water generally predominates the exposure to As, foods may be an important source and key risk factor. Furthermore, ingestion risk of As may further increase over time when irrigation with As-contaminated water continues while efforts to provide As-safe drinking water expand. Food is, however, not only a potential source of As; it is the foremost source of micronutrients. Micronutrient deficiency is prevalent in Bangladesh, with poor women and children suffering most. Therefore, the nutritional values of foods need to be taken into account when evaluating the levels of inorganic As in them. For example, when a vegetable is rich in both As and vitamin A, the nutritional value may have a greater health impact than its contribution to As exposure. The health risk assessment will require a number of challenging activities:

- Develop a food composition database, including seasonal and spatial variation. The currently available database is outdated and includes neither As nor micronutrients.
- Develop a food consumption database, including seasonal and spatial variation.
- Assess the dietary intake of As and nutrients.
- Develop a method to weigh the risks of As in foods against their nutritional content.
- Quantify the bioavailability of As and nutrients, taking malnutrition into account.
- Quantify the influence of food preparation on food composition and bioavailability.
- Evaluate existing guidelines and standards on As in water and food and the PMTDI, taking into account the country-specific conditions.
- Develop guidelines/standards on As in food to evaluate current and future levels of As in foods.

C. Increasing Quantity and Quality of Chemical Analyses

To overcome limitations in the quality of chemical analysis and research in general, the capacity of laboratories in Bangladesh needs to be strengthened. It is not only the chemical analysis itself that need to be improved, it is the entire range of procedures, data collection, and good laboratory practice from sampling in the field to data interpretation that determines the quality of the work. For example, basic issues related to trace-element analysis such as sample size, removal of potential sources of contamination such as dust from food samples, and homogenization are often addressed inappropriately or not at all. Methods should be validated and described in detail, and the development of As speciation capacity in Bangladesh is essential. Overall, internationally accepted principles on QA/QC should become the standard. One of the obstacles that laboratories in Bangladesh are facing are the costs of CRM. To overcome this, secondary reference materials can be developed relatively cheaply by a consortium of Bangladeshi institutes in collaboration with the international scientific community.

V. Conclusions

There are clearly two potential risks from As in irrigation water. The risk to food safety is publicly well known, but available data are still insufficient in terms of quality and quantity for a reliable and representative risk assessment. The risk to soil quality and crop yield is hardly known and has received very little attention. Our understanding of the long-term behaviour of As in agriculture in Bangladesh is too limited to assess the risks. It is clear that As is accumulating at various locations because of irrigation water input. There is also evidence that an increase in soil concentrations can be reflected by concentrations in crops, which probably depend on specific soil conditions and plant species and cultivars. These observations suggest that both an increase of As in crops and, at a later stage, As toxicity to crops cannot be excluded over time.

Summary

High levels of As in groundwater commonly found in Bangladesh and other parts of Asia not only pose a risk via drinking water consumption but also a risk in agricultural sustainability and food safety. This review attempts to provide an overview of current knowledge and gaps related to the assessment and management of these risks, including the behaviour of As in the soil–plant system, uptake, phytotoxicity, As speciation in foods, dietary habits, and human health risks. Special emphasis has been given to the situation in Bangladesh, where groundwater via shallow tube wells is the most important source of irrigation water in the dry season. Within the soil–plant system, there is a distinct difference in behaviour of As under flooded conditions, where arsenite (AsIII) predominates, and under nonflooded conditions, where arsenate (AsV) predominates. The former is regarded as most toxic to humans and plants. Limited data indicate that As-contaminated irrigation water can result in a slow buildup of As in the topsoil. In some cases the buildup is reflected by the As levels in crops, in others not. It is not yet possible to predict As uptake and toxicity in plants based on soil parameters. It is unknown under what conditions and in what time frame As is building up in the soil. Representative phytotoxicity data necessary to evaluate current and future soil concentrations are not yet available. Although there are no indications that crop production is currently inhibited by As, long-term risks are clearly present. Therefore, with concurrent assessments of the risks, management options to further prevent As accumulation in the topsoil should already have been explored. With regard to human health, data on As speciation in foods in combination with food consumption data are needed to assess dietary exposure, and these data should include spatial and seasonal variability. It is important to control confounding factors in assessing the risks. In a country where malnutrition is prevalent, levels of inorganic As in foods should be balanced against the nutritional value of the foods. Regarding agriculture, As is only one of the many factors that may pose a risk to the sustainability of crop production. Other risk factors such as nutrient depletion and loss of organic matter also must be taken into account to set priorities in terms of research, management, and overall strategy.

Acknowledgments

Sasha Koo-Oshima (Land and Water Division, FAO Headquarters), Thierry Facon, Zhijun Chen, Yuji Niino (FAO Regional Office for Asia and the Pacific), Han Heijnen and Roy Boerschke (WHO), Guy Howard (APSU), Paul Williams (University of Aberdeen), Henk Schat (Vrije Universiteit Amsterdam), Richard Loeppert (Texas A&M University), John Duxbury (Cornell University), Rob Ritsema (RIVM), and Rafiqul Islam (Bangladesh Agricultural University) are thanked for valuable comments and discussion. This study has been funded by the Government of the Netherlands via the Associate Professional Officer program.

References

Abedin, MJ, Meharg, AA (2002) Relative toxicity of arsenite and arsenate on germination and early seedling growth of rice (*Oryza sativa* L.). Plant Soil 243:57–66.

Abedin, MJ, Cotter-Howells, J, Meharg, AA (2002a) Arsenic uptake and accumulation in rice (*Oriza sativa* L.) irrigated with contaminated water. Plant Soil 240:311–319.

Abedin, MJ, Cressner, MS, Meharg, AA, Feldmann, J, Cotter-Howells, J (2002b) Arsenic accumulation and metabolism in rice (*Oryza sativa L.*). Environ Sci Technol 36:962–968.

Abedin, MJ, Feldmann, J, Meharg, AA (2002c) Uptake kinetics of arsenic species in rice plants. Plant Physiol 128:1120–1128.

Ahmad, SA, Bandaranayake, D, Khan, AW, Hadi, SA, Uddin, G, Halim, MA (1997) Arsenic contamination in groundwater and arsenicosis in Bangladesh. Int J Environ Health Res 7:271–276.

Ahmed, KM, Bhattacharya, P, Hasan, MA, Akhter, SH, Alam, SMM, Bhuyian, MAH, Imam, MB, Khan, AA, Sracek, O (2004) Arsenic enrichment in groundwater of the alluvial aquifers in Bangladesh: an overview. Appl Geochem 19:181–200.

Alam, MB, Sattar, MA (2000) Assessment of arsenic contamination in soils and waters in some areas of Bangladesh. Water Sci Technol 42:185–193.

Alam, MGM, Allinson, G, Stagnatti, F, Tanaka, A, Westbrooke, M (2002) Arsenic contamination in Bangladesh groundwater: a major environmental and social disaster. Int J Environ Health Res 12:236–253.

Alam, MGM, Snow, ET, Tanaka, A (2003) Arsenic and heavy metal contamination of vegetables grown in Samta village, Bangladesh. Sci Total Environ 308:83–96.

Ali, MA (2003) Fate of arsenic in the environment. In: Ahmed MF (ed) Arsenic Contamination: Bangladesh Perspective. ITN-Bangladesh, Dhaka, pp 84–100.

BADC (2004) Survey report on irrigation equipment and irrigated area in Boro/2003 season. pp 200. BADC, Dhaka.

Bae, M, Watanabe, C, Inaoka, T, Sekiyama, M, Sudo, N, Bokul, MH, Ohtsuka, R (2002) Arsenic in cooked rice in Bangladesh. Lancet 360:1839–1840.

Bleeker, PM, Schat, H, Vooijs, R, Verkleij, JAC, Ernst, WHO (2003) Mechanisms of arsenate tolerance in *Cytisus striatus*. New Phytol 157:33–38.

Bleeker, PM, Hakvoort, HWJ, Bliek, M, Souer, E, Schat, H (2006) Enhanced arsenate reduction by a CDC25-like tyrosine phosphatase explains increased phytochelatin accumulation in arsenate-tolerant Holcus lanatus. Plant J 45: 917–929.

Brammer, H (2005) The arsenic problem in Bangladesh. Tropical Agriculture Association Newsletter.

Breit, GN, Lowers, HA, Foster, AL, Perkins, RB, Yount, J, Whitney, JW, Uddin, MN, Muneem, AA (2005) Redistribution of arsenic and iron in shallow sediment of Bangladesh. In: Behavior of Arsenic in Aquifers, Soils and Plants. (Conference Proceedings), Dhaka, 2005.

Carbonell, AA, Aarabi, MA, DeLaune, RD, Gambrell, RP, Patrick, WH, Jr (1998) Arsenic in wetland vegetation: availability, phytotoxicity, uptake and effects on plant growth and nutrition. Sci Total Environ 217:189–199.

Carbonell-Barrachina, AA, Burló, F, Burgos-Hernández, A, López, E, Mataix, J (1997) The influence of arsenite concentrations on arsenic accumulation in tomato and bean plants. Sci Hortic 71:167–176.

Carbonell-Barrachina, AA, Aarabi, MA, Delaune, RD, Gambrell, RP, Patrick, WH, Jr (1998) The influence of arsenic chemical form and concentration on *Spartina putens* and *Spartina alterniflora* growth and tissue arsenic concentration. Plant Soil 198:33–43.

Carbonell-Barrachina, AA, Burló, F, Valero, D, López, E, Martínez-Romero, D, Martínez-Sánchez, F (1999) Arsenic toxicity and accumulation in turnip as affected by arsenic chemical speciation. J Agric Food Chem 74:2288–2294.

Cava-Montesinos, P, De la Guardia, A, Teutsch, C, Cervera, ML, De la Guardia, M (2003) Non-chromatographic speciation analysis of arsenic and antimony in milk hydride generation atomic fluorescence spectrometry. Anal Chim Acta 493: 195–203.

Chakraborti, D, Rahman, MM, Paul, K, Chowdhury, UK, Sengupta, MK, Lodh, D, Chanda, CR, Saha, KC, Mukherjee, SC (2002) Arsenic calamity in the Indian subcontinent. What lessons have been learned? Talanta 58:3–22.

Chen, Z, Zhu, Y-G, Liu, W-J, Meharg, AA (2005) Direct evidence showing the effect of root surface iron plaque on arsenite and arsenate uptake into rice (*Oryza sativa*). New Phytol 165:91–97.

Chowdhury, UK, Biswas, BK, Chowdhury, TR, Samanta, G, Mandal, BK, Basu, GC, Chanda, CR, Lodh, D, Saha, KC, Mukherjee, SK, Roy, S, Kabir, S, Quamruzzaman Q, Chakraborti, D (2000) Groundwater Arsenic Contamination in Bangladesh and West Bengal, India. Environ Health Perspect 108:393–397.

Cornu, S, Breeze, D, Saada, A, Baranger, P (2003) The influence of pH, electrolyte type, and surface coating on arsenic(V) adsorption onto kaolinites. Soil Sci Soc Am J 67:1127–1132.

Das, HK, Mitra, AK, Sengupta, PK, Hossain, A, Islam, F, Rabbani, GH (2004) Arsenic concentrations in rice, vegetables, and fish in Bangladesh: a preliminary study. Environ Int 30:383–387.

Dasgupta, T, Hossain, SA, Meharg, AA, Price, AH (2004) An arsenate tolerance gene on chromosome 6 of rice. New Phytol 163:45–49.

Dembitsky, VM, Rezanka, T (2003) Natural occurence of arseno compounds in plants, lichens, fungi, algal species, and microorganisms. Plant Sci 165:1177–1192.

Duxbury, JM, Zavala, YJ (2005) What are safe levels of arsenic in food and soils? In: Behavior of Arsenic in Aquifers, Soils and Plants. Conference Proceedings. International Symposium, Dhaka, 2005.

Duxbury, JM, Mayer, AB, Lauren, JG, Hassan, N (2003) Food chain aspects of arsenic contamination in Bangladesh: effects on quality and productivity of rice. J Environ Sci Health A38:61–69.

Farid, ATM, Sen, R, Haque, MA, Hossain, KM, Panaullah, GM, Meisner, CA, Loeppert, RH, Duxbury, JM (2005) Arsenic status of water, soil, rice grain and straw of individual shallow tube well command area of Brahmanbaria. In: Behavior of Arsenic in Aquifers, Soils and Plants. Conference Proceedings. International Symposium, Dhaka, 2005.

Fitz, WJ, Wenzel, WW (2002) Arsenic transformations in the soil-rhizosphere-plant system: fundamentals and potential application to phytoremediation. J Biotechnol 99:259–278.

Ghosh, AK, Bhattacharyya, P, Pal, R (2003) Effect of arsenic contamination on microbial biomass and its activities in arsenic contaminated soils of Gangetic West Bengal, India. Environ Int 30:491–499.

Hartley-Whitaker, J, Ainsworth, G, Vooijs, R, Ten, Bookum, WM, Schat, H, Meharg, AA (2001) Phytochelatins are involved in differential arsenate tolerance in *Holcus lanatus*. Plant Physiol 126:299–306.

Harvey, CF, Swartz, CH, Badruzzaman, ABM, Keon-Blute, N, Yu, W, Ashraf Ali, M, Jay, J, Beckie, R, Niedan, V, Brabander, DJ, Oates, PM, Ashfaque, KN, Islam, S, Hemond, HF, Ahmed, MF (2002) Arsenic mobility and groundwater extraction in Bangladesh. Science 298:1602–1606.

Heitkemper, DT, Vela, NP, Stewart, KR, Westphal, CS (2001) Determination of total and speciated arsenic in rice by ion chromatography and inductively coupled mass spectrometry. J Anal At Spectrom 16:299–306.

Hels, O, Hassan, N, Tetens, I, Thilsted, SH (2003) Food consumption, energy and nutrient intake and nutritional status in rural Bangladesh: changes from 1981–82 to 1995–96. Eur J Clin Nutr 57:586–594.

Hossain, MB (2005) Arsenic distribution in soil and water of a STW command area. In: Behavior of Arsenic in Aquifers, Soils and Plants. Conference Proceedings. Internal Symposium, Dhaka, 2005.

Huang, Y-K, Lin, K-H, Chen, H-W, Chang, C-C, Liu, C-W, Yang, M-H, Hsueh, Y-M (2003) Arsenic species contents at aquaculture farm and in farmed mouth-breeder (*Oreochromis mossambicus*) in blackfoot disease hyperendemic areas. Food Chem Toxicol 41:1491–1500.

Islam, FS, Gault, AG, Boothman, C, Polya, DA, Charnock, JM, Chatterjee, D, Lloyd, JR (2004a) Role of metal-reducing bacteria in arsenic release from Bengal delta sediment. Nature (Lond) 430:68–71.

Islam, MR, Islam, S, Jahiruddin, M, Islam, MA (2004b) Effects of irrigation water arsenic in the rice croppping system. J Biol Sci 4:542–546.

Islam, MR, Salminen, R, Lahermo, PW (2000) Arsenic and other toxic elemental contamination of groundwater, surface water and soil in Bangladesh and its possible effects on human health. Environ Geochem Health 22:33–53.

Islam, MR, Jahiruddin, M, Rahman, GKMM, Miah, MAM, Farid, ATM, Panaullah, GM, Loeppert, RH, Duxbury, JM, Meisner, CA (2005) Arsenic in paddy soils of Bangladesh: levels, distribution and contribution of irrigation and sediments. In: Behavior of Arsenic in Aquifers, Soils and Plants. Conference Proceedings. International Symposium, Dhaka, 2005.

Jahiruddin, M, Islam, MA, Islam, MR, Islam, S (2004) Effects of arsenic contamination on rice crop. Environtropica 1:204–210.

Jahiruddin, M, Islam, MR, Shah, MAL, Rashid, MA, Ghani, MA (2005) Arsenic in the water-soil-crop systems: PETRRA-BRRI-BAU-AAS study. In: Behavior of Arsenic in Aquifers, Soils and Plants. Conference Proceedings. International Symposium, Dhaka, 2005.

JICA/AAN (2004) Arsenic contamination of irrigation tubewells in Sharsha subdistrict, Jessore. Asia Arsenic Network, Dhaka.

Khan, MA (2005) Movement of arsenic in undisturbed soil columns with irrigated rice culture. In: Behavior of Arsenic in Aquifers, Soils and Plants. Conference Proceedings. International Symposium, Dhaka, 2005.

Kohlmeyer, U, Jantzen, E, Kuballa, J, Jakubik, S (2003) Benefits of high resolution IC-ICP-MS for the routine analysis of inorganic and organic arsenic species in food products of marine and terrestrial origin. Anal Bioanal Chem 377:6–13.

Lambkin, DC, Alloway, BJ (2003) Arsenate-induced phosphate relaese from soils and its effect on plant phosphorus. Water Air Soil Pollut 144:41–56.

Lamont, W (2003) Concentrations of inorganic arsenic in samples of white rice from the United States. J Food Comp Anal 16:687–695.

Lauren, JG, Duxbury, JM (2005) Management strategies to reduce arsenic uptake by rice. In: Behavior of Arsenic in Aquifers, Soils and Plants. Conference Proceedings. International Symposium, Dhaka, 2005.

Li, W, Wei, C, Zhang, C, Van, Hulle, M, Cornelis, R, Zhang, X (2003) A survey of arsenic species in chinese seafood. Food Chem Toxicol 41:1103–1110.

Liu, W-J, Zhu, Y-G, Smith, FA, Smith, SE (2004) Do phosphorus nutrition and iron plaque alter arsenate (As) uptake by rice seedlings in hydroponic culture? New Phytol 162:481–488.

Mahimairaja, S, Bolan, NS, Adriano, DC, Robinson, B (2005) Arsenic contamination and its risk management in complex environmental settings. Adv Agron 86:1–82.

Mandal, BK, Suzuki, KT (2002) Arsenic around the world: a review. Talanta 58:201–235.

Mandal, BK, Chowdhury, TR, Samanta, G, Mukherjee, DP, Chanda, CR, Saha, KC, Chakraborti, D (1998) Impact of safe water for drinking and cooking on five arsenic-affected families for 2 years in West Bengal, India. Sci Total Environ 218:185–201.

Masscheleyn, PH, Delaune, RD, Patrick, WH (1991) Effect of redox potential and pH on arsenic speciation and solubility in a contaminated soil. Environ Sci Technol 25:1414–1419.

Matschullat, J (2000) Arsenic in the geosphere: a review. Sci Total Environ 249:297–312.
McArthur, JM, Banjeree, DM, Hudson-Edwards, KA, Mishra, R, Purohit, R, Ravenscroft, P, Cronin, A, Howarth, RJ, Chatterjee, A, Talukder, T, Lowry, D, Houghton, S, Chadha, DK (2004) Natural organic matter in sedimentary basins and its relation to arsenic in anoxic ground water: the example of West Bengal and its worldwide implications. Appl Geochem 19:1255–1293.
Meharg, AA (2004) Arsenic in rice: understanding a new disaster for South-East Asia. Trends Plant Sci 9:415–417.
Meharg, AA, Hartley-Whitaker, J (2002) Arsenic uptake and metabolism in arsenic resistant and nonresistant plant species. New Phytol 154:29–43.
Meharg, AA, Jardine, L (2003) Arsenite transport into paddy rice (*Oryza sativa*) roots. New Phytol 157:39–44.
Meharg, AA, Rahman, MM (2003) Arsenic contamination of Bangladesh paddy field soils: implications for rice contribution to arsenic consumption. Environ Sci Technol 37:229–234.
Miah, MAM, Rahman, MS, Islam, A, Paul, DNR, Farid, ATM, Jahiruddin, M, Sattar, MA, Panaullah, GM, Meisner, CA, Loeppert, RH, Duxbury, JM (2005) Nation-wide survey of arsenic in soils, water and crops in Bangladesh. In: Behavior of Arsenic in Aquifers, Soils and Plants. Conference Proceedings. International Symposium, Dhaka, 2005.
Ng, JC, Wang, J, Shraim, A (2003) A global health problem caused by arsenic from natural sources. Chemosphere 52:1353–1359.
Nicolas, DR, Ramamoorthy, S, Palace, V, Spring, S, Moore, JN, Rosenzweig, RF (2003) Biogeochemical transformations of arsenic in circumneutral freshwater sediments. Biodegradation 14:123–137.
Norra, S, Berner, ZA, Agarwala, P, Wagner, F, Chandrasekharam, D, Stuben, D (2005) Impact of irrigation with As rich groundwater on soil and crops: a geo-chemical case study in West Bengal Delta Plain, India. Appl Geochem 20:1890–1906.
Onken, BM, Hossner, LR (1995) Plant uptake and determination of arsenic species in soil solution under flooded conditions. J Environ Qual 24:373–381.
Panaullah, GM, Ahmed, ZU, Duxbury, JM, Loeppert, RH, Meisner, CA (2005) Arsenic in the irrigation water–soil environment and flows of arsenic in the food system. In: Behavior of Arsenic in Aquifers, Soils and Plants. Conference Proceedings. International Symposium, Dhaka, 2005.
Pizarro, I, Gomez, M, Camara, C, Palacios, MA (2003a) Arsenic speciation in environmental and biological samples Extraction and stability studies. Anal Chim Acta 495:85–98.
Pizarro, I, Gomez, M, Palacios, MA, Camara, C (2003b) Evaluation of stability of arsenic species in rice. Anal Bioanal Chem 376:102–109.
Raab, A, Schat, H, Meharg, AA, Feldmann, J (2005) Uptake, translocation and transformation of arsenate and arsenite in sunflower (*Helianthus annuus*): formation of arsenic–phytochelatin complexes during exposure to high arsenic concentrations. New Phytol 168:551–558.
Ravenscroft, P, McArthur, JM, Hoque, BA (2001) Geochemical and palaeo-hydrological controls on pollution of groundwater by arsenic. In: Chappel WR, Abernathy CO, Calderon R (eds). Arsenic Exposure and Health Effects, vol IV. Elsevier, Oxford, pp 53–78.

Ross, Z, Duxbury, JM, Paul, DNR, DeGloria, SD (2005) Potential for arsenic contamination of rice in Bangladesh: spatial analysis and mapping of high risk areas. In: Behavior of Arsenic in Aquifers, Soils and Plants. Conference Proceedings. International Symposium, Dhaka, 2005.

Roychowdhury, T, Uchino, T, Tokunaga, H, Ando, M (2002a) Arsenic and other heavy metals in soils from an arsenic-affected area of West Bengal, India. Chemosphere 49:605–618.

Roychowdhury, T, Uchino, T, Tokunaga, H, Ando, M (2002b) Survey of arsenic in food composites from an arsenic affected area of West Bengal, India. Food Chem Toxicol 40:1611–1621.

Sanz, E, Munoz-Olivas, R, Camara, C (2005) A rapid and novel alternative to conventional sample treatment for arsenic speciation in rice using enzymatic ultrasonic probe. Anal Chim Acta 535:227–235.

Schmidt, A-C, Mattusch, J, Reisser, W, Wennrich, R (2004) Uptake and accumulation behaviour of angiosperms irrigated with solutions of different arsenic species. Chemosphere 56:305–313.

Schoof, RA, Yost, LJ, Crecelius, EA, Cragin, DW, Meacher, DM, Menzel, DB (1999) A market basket survey of inorganic arsenic in food. Food Chem Toxicol 37:839–846.

Smedley, PL, Kinniburgh, DG (2002) A review of the source, behaviour and distribution of arsenic in natural waters. Appl Geochem 17:517–568.

Sneller, FEC, Noordover, ECM, Ten, Bookum, WM, Schat, H, Bedaux, JJM, Verkleij, JAC (1999a) Quantitative relationship between phytochelatin accumulation and growth inhibition during prolonged exposure to cadmium in *Silene vulgaris*. Ecotoxicology 8:167–175.

Sneller, FEC, Van, Heerwaarden, LM, Kraaijeveld-Smit, FJL, Ten, Bookum, WM, Koevoets, PLM, Schat, H, Verkleij, JAC (1999b) Toxicity of arsenate in *Silene vulgaris*, accumulation and degradation of arsenate-induced phytochelatins. New Phytol 144:223–232.

Sneller, FEC, Van, Heerwaarden, LM, Schat, H, Verkleij, JAC (2000) Toxicity, metal uptake, and accumulation of phytochelatins in *Silene vulgaris* exposed to mixtures of cadmium and arsenate. EnvironToxicol Chem 19:2982–2986.

Takahashi, Y, Minamikawa, R, Hattori, KH, Kurishima, K, Kihou, N, Yuita, K (2004) Arsenic behavior in paddy fields during the cycle of flooded and non-flooded periods. Environ Sci Technol 38:1038–1044.

Watanabe, C, Kawata, A, Sudo, N, Sekiyama, M, Inaoka, T, Bae, M, Ohtsuka, R (2004) Water intake in an Asian population living in arsenic-contaminated area. Toxicol Appl Pharmacol 198:272–282.

Weiss, JV, Emerson, D, Backer, SM, Megonigal, JP (2003) Enumeration of Fe(II)-oxidizing and Fe(III)-reducing bacteria in the root zone of wetland plants: implications for a rhizosphere iron cycle. Biogeochemistry 64:77–96.

Weiss, JV, Emerson, D, Megonigal, JP (2004) Geochemical control of microbial Fe(III) reduction potential in wetlands: comparison of the rhizosphere to nonrhizospere soil. FEMS Microbiol Ecol 48:89–100.

WHO (1996) Trace elements in human nutrition and health. WHO, Geneva.

WHO (2001) Environmental Health Criteria 224: arsenic and arsenic compounds. WHO, Geneva.

Williams, LE, Barnett, MO, Kramer, TA, Melville, JG (2003) Adsorption and transport of arsenic(V) in experimental subsurface systems. J Environ Qual 32: 841–850.

Williams, PN, Price, AH, Raab, A, Hossain, SA, Feldmann, J, Meharg, AA (2005) Variation in arsenic speciation and concentration in paddy rice related to dietary exposure. Environ Sci Technol 39:5531–5540.

Yamazaki, C, Ishiga, H, Ahmed, F, Itoh, K, Suyama, K, Yamamoto, H (2003) Vertical distribution of arsenic in Ganges delta sediments in Deuli Village, Bangladesh. Soil Sci Plant Nutr 49:567–574.

Zheng, Y, Stute, M, van Geen, A, Gavrieli, I, Dhar, R, Simpson, HJ, Schlosser, P, Ahmed, KM (2004) Redox control of arsenic mobilization in Bangladesh groundwater. Appl Geochem 19:201–214.

Zobrist, J, Dowdle, PR, Davis, JA, Oremland, RS (2000) Mobilization of arsenite by dissimilatory reduction of adsorbed arsenate. Environ Sci Technol 34:4747–4753.

Manuscript received March 8; accepted March 14, 2006.

Health Effects of Arsenic, Fluorine, and Selenium from Indoor Burning of Chinese Coal

Liu Guijian, Zheng Liugen, Nurdan S. Duzgoren-Aydin, Gao Lianfen, Liu Junhua, and Peng Zicheng

Contents

I. Introduction 89
II. Arsenic 92
III. Arsenic Epidemic in China 94
IV. Fluorine 96
V. Fluorine Epidemic in China 96
VI. Selenium 98
VII. Selenium Epidemic in China 98
VIII. Conclusions 99
Summary 100
Acknowledgments 100
References 100

I. Introduction

China's economy has developed rapidly in the last two decades, leading to an increase in energy consumption and consequently emissions from energy generation. China is the largest coal producer and consumer in the world (Finkelman 1995). It has been estimated that more than 75% of the energy production in China is based on coal (Chen et al. 2004), and more than 400 million people in China rely on coal for their domestic energy needs, such as heating and daily cooking. Due to the limited petroleum and natural gas reserves and significant coal reserves (1 trillion t) in China, it is likely that this coal-based, relatively cheap energy structure will continue for the foreseeable future (Ni 2000; Xu et al. 2000; Zhong and Wang 2000).

Communicated by David M. Whitacre.

Liu GJ (✉) · Zheng LG · Gao LF · Liu JH · Peng ZC
CAS Key Laboratory of Crust-Mantle Materials and Environments, School of Earth and Space Sciences, University of Science and Technology of China, Hefei Anhui 230026, China

Liu GJ · Peng ZC
State Key Laboratory of Loess and Quaternary Geology, Institute of Earth Environment, The Chinese Academy of Sciences, Xi'an 710075, Shaanxi, P.R. China

N.S. Duzgoren-Aydin
Department of Earth Sciences, The University of HongKong, Pokfulam Road, Hong Kong, SAR China

Coal is a complex heterogeneous mixture of organic and inorganic constituents of allothigenic or authigenic origin. Besides major (>1%) and minor (0.1%–1%) elements in coal, elements such as As, Ba, Se, Hg, Cr, Mn, Mo, Sb, Th, U, and V occur commonly as trace elements (<1,000 ppm) associated with both organic and inorganic matter. Trace elements in coals are the major pollution sources (Swaine 2000), and they may become easily accessible not only during coal mining, but also during the storage, transport, cleaning, combustion, and other coal preparation processes to make the coal consistent in quality and suitable for selling (Ni 2000; Liu and Peng 2002a).

Trace elements in coals can disperse into the environment through several pathways, such as leaching into soil and adversely affecting plant nutrition and by creating runoff into waterways. More importantly, some trace elements, particularly those easily volatilized during combustion (e.g., As, F, and Se), release into the atmosphere either as vapor or respirable fine particulates (Liu et al. 2004b). Therefore, these toxic elements can easily enter into the human body through the respiratory tract or by ingestion and accumulate in tissues. When the concentrations of these trace elements are higher than the required intake by humans, they will adversely affect normal metabolism, interfere with physiological functions, and can produce human pathology. These health problems can be severe, leading to death, and can be widespread, affecting many millions of people.

Ezatti et al. (2002) estimated that in 2000, global mortality caused by indoor air pollution from solid fuels (such as wood, charcoal, and crop residues, but mainly coal), was more than 1.6 million. This rate of severe global mortality together with combustion-induced diseases such as pneumoconiosis, dermatosis, and other relevant diseases indicate an urgent need to take immediate action against indoor air pollution, especially in developing regions. This concern is particularly valid in China where elemental toxicosis is epidemic (Zheng et al. 1996, 1999). The World Bank (1992) estimates that between 400 and 700 million people, particularly women and children, are exposed to severe indoor air pollution in China. A substantial proportion of these people rely on coal for domestic cooking and heating, gases, and organic compounds. In particular, significant numbers of people in Southwest China have been exposed to high levels of toxic trace elements emitted from coal combustion for many years. This chronic poisoning due to burning of toxic element-bearing coals has been further exacerbated by consumption of foodstuffs prepared by drying directly over coal fires (Fig. 1) (Zheng 1992; Zheng and Huang 1984, 1985, 1986, 1989).

Initially a total of 11 trace elements emitted during coal combustion were identified as hazardous air pollutants (HAPs) (U.S. National Committee for Geochemistry 1990). Recently, this number has been increased to 26

Fig. 1. Indoor stoves without chimneys.

Table 1. Trace elements considered to be of environmental interest

Groups	Trace elements
1	As, Cd, Cr, Hg, Pb, Se
2	B, Cl, F, Mn, Mo, Ni, Be, Cu, Th, P, U, V, Zn
3	Ba, Co, I, Ra, Sb, Sn, Tl

(Swaine 2000). These 26 HAPs (Table 1) are considered to be of environmental concern and considered as primary targets for the implementation of stringent regulations to protect human health worldwide. Although the environmental importance of the elements decreases from group 1 to group 3 in Table 1, we keep in mind that every element, with the exception of As, may be both essential and hazardous. Detailed knowledge of toxic trace elements in coals, therefore, is required to understand their behavior during coal processing and utilization and consequential environmental impacts

(Finkleman 1981, 1989). In this review we focus on the geochemistry of As, Se, and F in Chinese coal, document their concentrations, distributions, modes of occurrence, and volatility during coal combustion, and then discuss As, Se, and F toxicosis in Southwest China (Guizhou and Hubei, respectively).

II. Arsenic

Arsenic (As), ubiquitous in the environment, is considered one of the most hazardous elements because of its carcinogenic potential. Arsenic can enter the body through inhalation, ingestion, and skin absorption. Once absorbed, arsenic can distribute to various parts of the human body and accumulate in hair and nails, causing chronic arsenic poisoning (arseniasis) (An et al. 1992; Zhou et al. 1993; Belkin et al. 1997; Zheng et al. 1999; Li et al. 2000; Finkleman et al. 1999, 2002; Zhang et al. 1999; Zhou et al. 2002; Huang et al. 2002). One of the major sources of arsenic contamination (the highly toxic oxides such as As_2O_3 and As_2O_5 in particular) comes from the combustion of arsenic-containing coal (Belkin et al. 1997; Zheng et al. 1999; Finkleman et al. 1999, 2002).

The concentration and distribution of As in Chinese coals has been extensively investigated since 1980 (Cui et al. 1998; Liu et al. 1999a,b; Liu et al. 2001a,b, 2002b, 2004a; Li et al. 2002; He et al. 2002; Ding et al. 2001; Zhang et al. 2004). In general, the concentration of As in Chinese coals is similar to the global average value, which is around 5 ppm. However, depending upon the geochemical and geological characteristics of the particular coal, arsenic concentrations can be variable and change significantly within different regions. For example, in localities such as Guilin, Zhejiang, Fujian, Hubei, Hunan, Guangdong, and Guangxi provinces (Fig. 2), the average arsenic concentration in the coal is about 10 ppm, whereas in Southwest China, including Yunan, Guizhou, Sichuan, and Chongqing provinces. As concentrations can vary from 0.1 to 1,483 ppm, with an average value of 18.16 ppm (Li et al. 2002).

The chemical state and mode of occurrence of an element (Zhao and Zhang 1988) determine its toxicity and behavior, such as potential release during mining, processing, and utilization. Detailed knowledge about the mode of occurrence of an element is, therefore, crucial to assess its environmental impact (Finkelman 1989), and numerous studies have been made of the mode of occurrence of arsenic in coal (Finkelman 1995; Belkin et al. 1997; Zhou 1998; Zhao et al. 1998, 2003; Ding et al. 1999; Kolker et al. 2000; Huang et al. 2000; Liu et al. 2002b; Chen et al. 2002; Guo et al. 2003; Liu et al. 2004a). The general consensus of these studies are as follows: (a) As_2O_3 and As_2O_5 as the most common and most toxic forms of inorganic arsenic, being most often found in As-bearing minerals such as realgar (AsS), orpiment (As_2S_3), and arsenopyrite (FeAsS); (b) arsenic is

Fig. 2. Distribution of arsenic in Chinese coals.

also found in association with sulfides of copper, lead, and zinc and with the minerals pyrite, chalcopyrite, and sphalerite; (c) approximately 80% arsenic in coal occurs as an inorganic compound (Finkelman 1995); and (d) arsenic in coal occurring as an organic compound can be totally volatilized during combustion (Finkelman 1981; Ding et al. 1999, 2003; Liu et al. 2002a).

Arsenic has a melting point of 1,090 K and a sublimation temperature of 889 K, and vaporizes during coal combustion. Although As content in the majority of the coals is relatively low, the amount of As released during coal combustion is significant because of the large amount of coal used over extended periods. For instance, about 117×10^9 t coal was consumed worldwide during 1900–1971, resulting in the release of 270,000 t As to the environment (Swaine and Goodarz 1995). In China, about 5.45×10^9 t coal was consumed in 1980, and more than 5,000 t As was introduced into the environment every year (Wang et al. 1992). As released during coal combustion will be transported long distances and, therefore, can impact not only the air, soil, and biota both near and close to the area of combustion but also in remote areas (Swaine 2000; Liu et al. 2002a,b, 2003; Liu 2004). For example, Nerin et al. (1994) analyzed the As content in 38 clay samples surrounding a large coal-fired power plant in Spain and found significant As contamination in soils 80 km away. In addition to coal combustion, other processes such as coal mining, storage, and transport may also transform (e.g., via redox reactions) As-bearing compounds and introduce As into the environment (Feng et al. 1999; Liu et al. 2001a, 2004c).

Many researchers (Clarke 1993; Linak and Wendt-Jast 1994; Rizeq et al. 1994; Helble 1994; Swaine 1994; Querol et al. 1995; Constance 2000; Mastral et al. 2000; Edward 2000; Liu et al. 2004c) have shown that As volatility is higher during combustion and changes with temperature. Liu et al. (2004c), Guo et al. (1994), and Sun et al. (1984) documented that more than 25% of the As vaporized at temperatures above 800°C, and As vaporization increased to more than 60% as the temperature increased to 1,100°C. As vapor generated during coal combustion is often associated with fine particulates and is introduced to the atmosphere with these aerosols. Yang and Qian (1983) analyzed the combustion products from a coal-fired power plant in Tianjin and found that, although the As content in the coal was 6.5 ppm, the As content in bottom ash was reduced to 3.8 ppm and that in fly ash was enriched to 88 ppm. Likewise, Liu et al. (2004b) observed that while the As content was only 2.31 ppm in the coal used in a coal-fired power plant in the Yianzhou area, the As content in fly ash was three times higher (6.95 ppm). Furthermore, As compounds in coal undergo transformations during combustion and may form the poisonous oxides As_2O_3 (arsenic trioxide) and As_2O_5 (arsenic pentoxide), which are extremely hazardous (Lu et al. 1995; Lu et al. 1996; Xu and Zhai 1997; Liu et al. 2004d). To summarize, As is one of the most hazardous air pollutants causing severe health problems globally.

III. Arsenic Epidemic in China

Several cities in China, including Fushun, Shenyang, Lanzhou, Guiyang, Chengdou, and Chongqing, not only suffer from poor air quality but also from significant As air pollution, mainly because of the burning of high-arsenic-bearing coals (Liu et al. 2001a,b; Chen et al. 2002, 2004). Chronic As poisoning in these regions has been reported for more than three decades. Currently, there are more than 33,000 people in Guizhou province alone affected by As pollution caused by burning high-As coals in poorly ventilated areas (Zheng et al. 1996, 1999; Li et al. 2000; Belkin et al. 1997; Finkelman et al. 1999, 2002). The ratio of people dying of liver cancer and hepatocirrhosis in these areas is also much higher than in other areas of China. More specifically, about 877 local people were diagnosed with As poisoning in 1975–1976 in Jiaole, Xingyi of Guizhou province, and the As concentration in coals in this area is as high as 9,600 ppm (Zheng et al. 1999). Ke (1980) reported chronic arsenic poisoning in Kaiyang, Guizhou, where the As content of coal is around 100 ppm. Similarly, Sun et al. (1984) reported the As concentration in coal samples to be 7,180 ppm in Zhijin, Guiyang, where 75 people were diagnosed as having As poisoning in 1964–1965. An et al. (1992) also confirmed that the average concentration in coals used for domestic heating is 2,170 ppm in Guizhou. Zhou et al. (1993) investigated the same area in Guizhou and found the highest arsenic in coals to be 8,300 ppm; they also reported that 1,548 people suffered from serious As poisoning. At the same time, researchers have found similar cases of As poisoning in other areas in Guizhou. Nevertheless, there are still at least 333,905 people in Guizhou province alone affected by As pollution due to the use of high-As coals.

In 2003, the authors in collaboration with Prof. Baoshan Zheng went to Zhijin, Guizhou, to investigate the pathology of As poisoning. The dominant symptoms of As poisoning are change of skin pigments and the increase of cutin cells of the palm, in addition to inflammation of the nervous and digestive systems. The organic injury is mainly reflected by nonvisible hepatomegaly, whereas skin injury is the most obvious significant symptom and includes hyperpigmentation (flushed appearance, freckles), hyperkeratosis (scaly lesions on the skin, generally concentrated on the hands and feet), Bowen's disease (dark, horny, precancerous lesions of the skin), and squamous cell carcinoma (Fig. 3).

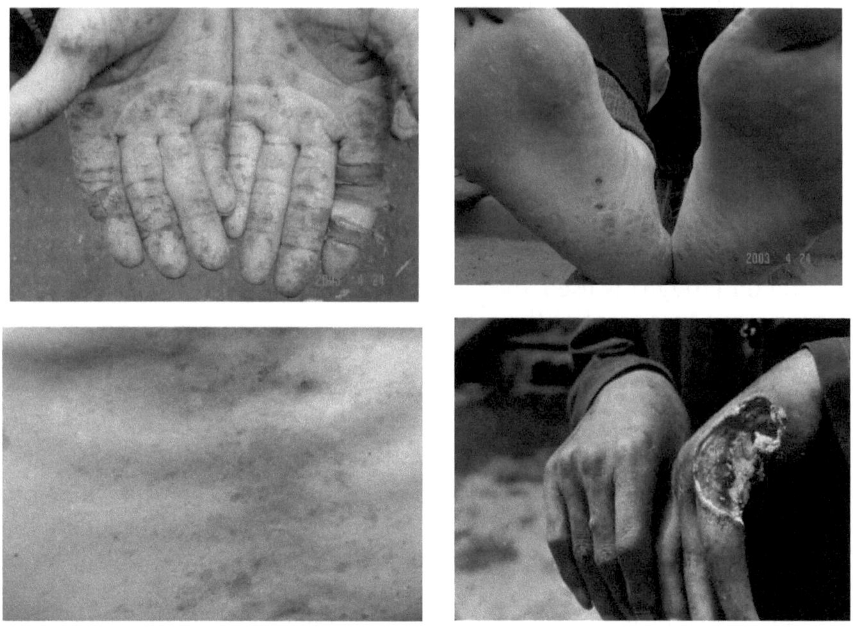

Fig. 3. Arsenism caused by burning coal indoors.

IV. Fluorine

Fluorine (F) is one of the most toxic and volatilized elements present in coal. During combustion, F is emitted as HF, SiF_4, and CF_4 (Lu 1996; Jeng et al. 1998; Yan et al. 1999; Liu et al. 1999a,b; Luo et al. 2001; Notcutt and Davies 2001). HF is one of the most serious pollutants adversely affecting both plants and animals, the toxicity being about 10–100 times higher than that of SO_2 (Jeng et al. 1998; Piekos and Paslawska 1999; Notcutt and Davies 2001). Much research has been carried out in China on its occurrence (Zheng and Cai 1988; Lu 1996; Liu et al. 1999a,b, 2000; Ren et al. 1999; Luo et al. 2001; Chen and Tang 2002). Zheng et al. (1988) first reported the average F content in Chinese coals as 200 mg/kg, much more than the world average of 80 mg/kg (Swaine 1990). Luo et al. (2004) collected 300 Chinese coal samples to study the F content and distribution. The results showed that the F concentration in most Chinese coals is less than 200 mg/kg, but it is higher than the global average concentration of 80 mg/kg, especially in Guizhou, Hubei, Yunan, and Hunan provinces, where F content in coal is as high as 3,000 mg/kg.

The mode of occurrence of F in coals varies significantly. Fluorine may occur as inorganic as well as organic compounds (Godbeer and Swaine 1987; Swaine 1990; Finkelman 1994; Liu et al. 1999b; Zheng et al. 1999; Liu et al. 2000, 2002b). The U.S. National Environmental Trust reported that the F emission is about 8% of total power plant pollutant emissions, following HCl (69%) and H_2SO_4 (21%), and becomes the third most important pollutant. It has been shown that about 85%–90% of the F present in coal is emitted as HF, either as vapor or absorbed on fine ash particles when the temperature is above 850°C (Guo et al. 1994; Liu et al. 1999a,b, 2004d).

The air, water, and soil around the Huaibei coal-burning power plant have all been affected by F pollution to varying degrees, the power plant having been identified as the source of the emissions (Tang et al. 1999a,b). Luo et al. (2002) measured the annual coal consumption in China for power plants and domestic heating as about 8×10^9 t. If the lower range of F concentration (100 g/t) is used, the atmosphere contains about 66,000 t, which is more than twice that of the U.S.

V. Fluorine Epidemic in China

The health problems caused by F released during coal combustion are more extensive than those caused by As. More than 14 provinces (including the northwest part of Guizhou, the south part of Sichuan and northeast of Yunnan, the southwest part of Hubei, the outheast of Sichuan, and the northwest part of Hubei) and more than 30 million people in China suffer from various forms of fluorosis, and about 15 million people have been diagnosed as having fluorosis (Zheng et al. 1989; Zhang and Cao 1996; Finkelman et al. 1999, 2002). More specifically, about 10 million people in Guizhou Province and surrounding areas suffer from various forms of fluorosis. In Beimen Zheng and Huachun in Guizhou, about 78% of the inhabitants are diagnosed as having skeletal fluorosis, including osteosclerosis. Almost every family in this district has members suffering from skeletal fluorosis with limited movement of the joints and outward manifestations such as knock-knees, bowlegs, and spinal curvature. In Xiaotang Zheng, Pengshui, Sichuan province, among 5,633 residents, about 98% suffer from tooth enamel mottling (dental fluorosis). Zheng et al. (1999) and Finkelman et al. (1999) documented that 97% of people older than 8 yr suffer from the devastating dental fluorosis.

In 2003, the authors carried out a series of fluorosis surveys in Guizhou. In a survey of one elementary school, among 57 students, only 1 student was found to be free of dental fluorosis, whereas 99% of the student population was diagnosed as having dental fluorosis. During these surveys, the authors also found that the youngest patient was about 1 yr old. Typical signs of fluorosis include mottling of tooth enamel (dental fluorosis) and various forms of skeletal fluorosis (Figs. 4, and 5).

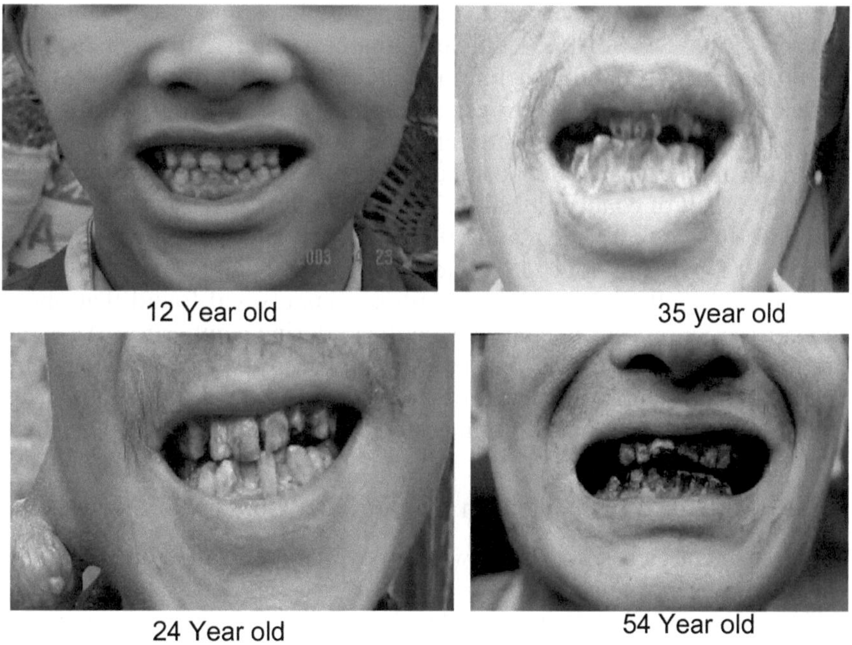

Fig. 4. Dental fluorosis caused by burning coal indoors.

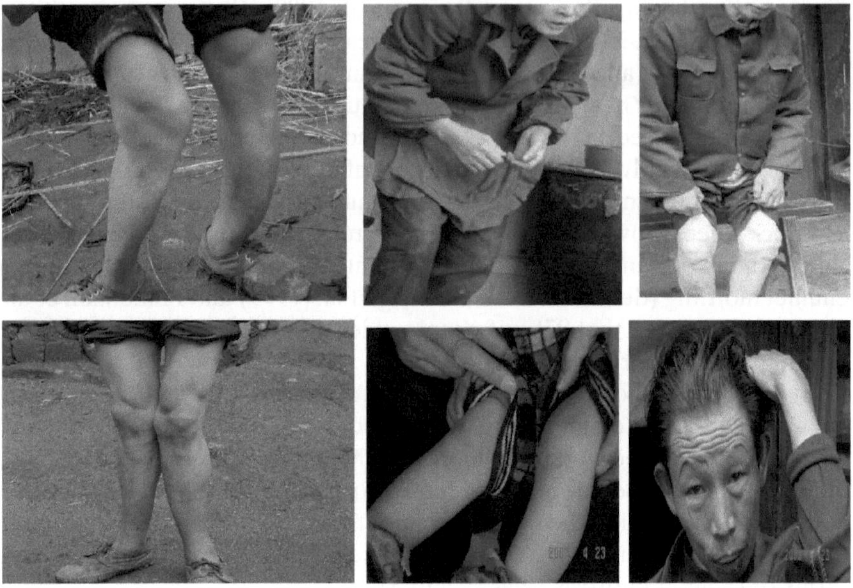

Fig. 5. Bone fluorosis caused by burning coal indoors.

VI. Selenium

Selenium (Se) is a necessary trace element for plants, animals, and humans. Se is present as Se^0, Se^{2-}, Se^{4+}, and Se^{6+} in natural waters. Although Se metal has little toxicity, selenide, selenite, and other Se compounds such as Se fluoride have a high toxicity. During coal combustion, Se is oxidized into SeO_2, which can easily dissolve in water and transform into selenite.

Nriagu and Pacyna (1988) postulated that atmospheric Se concentrations were the result of coal combustion. The results show the power industry is emitting $108–775 \times 10^7$ kg/yr, and other industrial and domestic use is $792–1,980 \times 10^7$ kg/yr. Sun et al. (1984) and Wang et al. (1996) carried out high-temperature (750°C) coal combustion experiments and found that about 75% of the Se present in coals is volatilized during combustion. Also, the ratio becomes higher (80%) in domestic heating appliances. During combustion, Se has similar characteristics to those of As and F, in that it is easily volatilized and can be inhaled or ingested, which may lead to Se poisoning.

VII. Selenium Epidemic in China

The first selenosis case in China was discovered at Yixing, Hubei in 1963. Nineteen of 23 residents in Yutang village were diagnosed as having Se poisoning, the symptoms being hair and nail loss, sometimes in association with disorders of the neural system. The inhabitants of the village had to be relocated. Similarly, in the other five villages of Yutang, about 49% of the res-

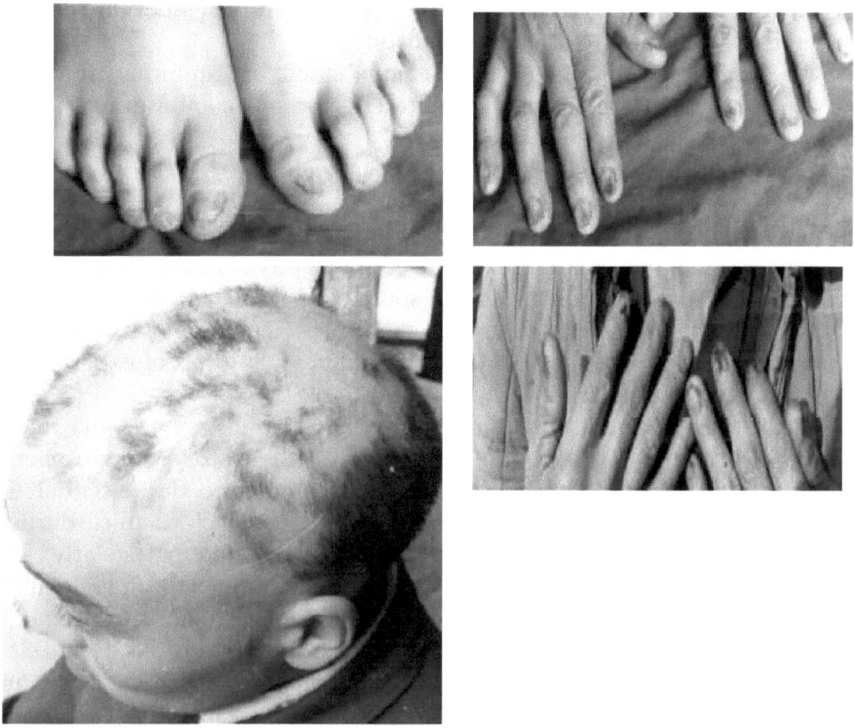

Fig. 6. Epidemic selenosis caused by burning coal indoors.

idents suffered from selenosis (Yang et al. 1981). Su et al. (1990) reported that there were about 477 regional selenosis cases from 1923 to 1987 in Yixin, Hubei, alone. Widespread symptoms were also identified in other parts of China, such as Ziyang, Shanxi (Cheng 1980). Unfortunately, the residents of Hubei are still using high-Se coals for heating and cooking. In 2003, the authors visited residents in Enshi, Xuanen, and Badong cities of Hubei province and encountered many selenosis cases (Fig. 6). In this area, the atmospheric Se concentration is as high as 0.6 mg/m^3 during coal combustion, with the highest value being 1.2 mg/m^3, resulting in several selenosis cases.

VIII. Conclusions

1. Arsenic, fluorine, and selenium are highly volatile toxic elements in Chinese coals, and their content tends to be higher in Chinese coals, especially in Southwest China.
2. Domestic coal containing high levels of these potentially toxic elements are burned indoors in open stoves, and foodstuffs are dried over these

stoves in some areas, causing serious health problems because of the emissions absorbed onto these foodstuffs.

3. In Southwest China, many of the inhabitants are suffering from the effects of toxic trace elements emitted during coal combustion.

Summary

China's economy has developed rapidly in the last two decades, leading to an increase in energy consumption and consequently emissions from energy generation. Coal is a primary energy source in China because of its abundance and will continue to be used in the future. The dominance of coal in energy production is expected to result in increasing levels of exposure to environmental pollution in China.

Toxic trace elements emitted during coal combustion are the main sources of indoor air pollution. They are released into the atmosphere mainly in the forms of fine ash and vapors and have the potential to adversely affect human health. Those trace elements, which volatilize during combustion, are hazardous air pollutants (HAPs) and are particularly rich in Chinese coals. Among the HAPs, arsenic (As), fluorine (F), and selenium (Se) have already been identified as pollutants that can induce severe health problems. In this review, the geochemical characteristics of As, F, and Se, including their concentration, distribution, and mode of occurrences in Chinese coal, are documented and discussed. Our investigations have confirmed the current As- and F-induced epidemics in Guizhou (Southwest China) and Se epidemic in Hubei (Northeast China). In this study, diagnostic symptoms of arseniasis, fluorosis, and selenosis are also illustrated.

Acknowledgments

This work was supported by National Natural Science Foundation of China (40133010, 40273035) and Anhui Natural Science Excellent Youth Foundation (04045064). Thanks are due to Professors Zheng Baoshan and Tang Yaogang for helping in the field studies.

References

An, D, He, G, Hu, X (1992) Combined intoxication of arsenic and fluorine caused by smoke pollution (in Chinese with English abstract). Guizhou Med 16(1):1–5.

Belkin, HE, Zheng, B, Zhou, D (1997) Preliminary results on the geochemistry and mineralogy of arsenic in mineralized coals from endemic arsenosis area in Guizhou province, China. In: Proceedings of the 14th Annual International Pittsburgh Coal Conference Workshop. CD-ROM ISBN 1-890977-14-4 (recorded in USA).

Chen, P, Kuang, H, Tang, X (2002) Research on the distribution and occurrence of arsenic in coal (in Chinese with English abstract). J China Coal Soc 27(3): 259–263.
Chen, P, Tang, X (2002) Fluorine in coal of China (in Chinese with English abstract). Coal Geol China 14(suppl):25–28.
Chen, B, Hong, C, Kan, H (2004) Exposures and health outcomes from outdoor air pollutants in China. Toxicology 198:291–300.
Cheng, J (1980) The preliminary investigation report of selenium intoxication in Ziyang County of Shanxi Province (in Chinese with English abstract). Shanxi Agric Sci 6:17.
Clarke, LB (1993) The fate of trace elements during coal combustion and gasification: an overview. Fuel 72(6):731–736.
Constance, LS (2000) Development of a mechanistic model for prediction of emission of trace elements from coal fired. Fuel Process Technol 63:75–77.
Cui, F, Chen, H (1998) Distribution and modes of occurrence of arsenic in coal of China (in Chinese with English abstract). Coal Sci Technol 26(12):44–46.
Ding, Z, Zheng, B, Zhan, J (1999) Preliminary study on the mode of occurrence of arsenic in high arsenic coals from southwest Guizhou province. Sci China 42(6):655–661.
Ding, Z, Zheng, B, Long, J, et al. (2001) Geological and geochemical characteristics of high arsenic coals from endemic arsenosis areas in southwestern Guizhou province, China. Appl Geochem 16:1353–1360.
Ding, Z, Zheng, B, Finkelman, RB (2003) Subsequent leaching study of typical high-As coal samples from southwest Guizhou province (in Chinese with English abstract). Earth Science – J China Univ Geosci 28(2):209–213.
Edward, F (2000) Characterization of trace element emission from coal combustion by equilibrium calculations. Fuel Process Technol 63:29–44.
Ezzati, M, Lopez, AD, Rodgers, A, Vander Hoorn, S, Murray, CJ (2002) Comparative risk assessment collaborative group of elected major risk factors and global and regional burden of disease. Lancet 360:1347–1360.
Fan, J, Zhang, Z (1995) Dynamic of trace elements during coal combustion (in Chinese with English abstract). Coal Process Compreh Util 4:12–14.
Feng, X, Hong, B, Ni, J, et al. (1999) Chemical mobility of potentially toxic trace elements in coal at surface conditions (in Chinese with English abstract). Acta Sci Circum 19(4):433–437.
Finkelman, RB (1981) Modes of Occurrence of Trance Elements in Coal Open-file rep 81–99. U S. Geology Survey, Washington, DC.
Finkelman, RB (1989) What we don't know about the occurrence and distribution of trace elements in coals. J Coal Qual 8:3–4.
Finkelman, RB (1994) Modes of occurrence of potentially hazardous elements in coal: levels of confidence. Fuel Process Technol 39:21–34.
Finkelman, RB (1995) Modes of occurrence of environmentally sensitive trace elements in coal. In: Swaine DJ, Goodarzi F (eds) Environmental Aspects of Trace Elements in Coal. Kluwer, Dordrecht, pp 24–50.
Finkelman, RB, Belkin, H, Zheng, B (1999) Health impacts of domestic coal use in China. Proc Natl Acad Sci USA 76:3427–3431.
Finkelman, RB, Orem, W, Castranova, V (2002) Health impacts of coal and coal use: possible solutions. Int J Coal Geol 50:425–443.
Godbeer, WC, Swaine, DJ (1987) Fluorine in Australian coals. Fuel 66:794–798.

Guo, X, Zheng, C, Liu, Y, Sun, T (2003) The behavior of mercury, arsenic, selenium during coal combustion (in Chinese with English abstract). J Eng Thermophys 22(6):763–766.

Guo, Y, Hou, H, Li, J, Chang, J (1994) Emission of arsenic, fluorine mercury, plumbium, cadmium during combustion process (in Chinese with English abstract). Coal Geol China 6(4):54–56.

He, B, Liang, L, Jiang, G (2002) Distribution of arsenic and selenium in selected Chinese coal mines. Sci Total Environ 296:19–26.

Helble, JJ (1994) Trace element behavior during coal combustion: results of a laboratory study. Fuel Process Technol 39:159–172.

Huang, W, Yang, Q, Tang, D, et al. (2000) Trace elements geochemistry of the coals in the Taiyuan formation from Zaozhuan Coal field. Geoscience 14(1):61–68.

Huang, X, Zhang, A, Yang, D, Hong, F (2002) Clinical sign and poly-organic and system damages of arensism caused by burning coal pollution and its sign ificance (in Chinese with English abstract). Chin J Endemiol 21(6):490–493.

Jeng, JH, Hsieh, CC, Lan, WH, et al. (1998) Cytotoxicity of sodium fluoride on human oral mucosal fibroblasts and its mechanisms. Cell Biol Toxicol 14(6):383–389.

Jin, Y, Liang, C, He, G, Cao, J, Ma, F (2003) Study on distribution of endemic arsenism in China. J Hyg Res 32(6):519–539.

Ke, C (1980) Investigation report of arsenism caused by domestic coal (in Chinese with English abstract). J Environ Sci 4(10).

Kolker, A, Huggins, FE, Palmer, CA, et al. (2000) Mode of occurrence of arsenic in four us coals. Fuel Process Technol 63:167–168.

Li, D, Chen, K, Deng, T, et al. (2002) Distribution and enrichment factors of arsenic in coal of southwestern China (in Chinese with English abstract). J China Univ Mining Technol 31(4):419–422.

Li, P, Liu, M, Li, D, Wang, S (2000) Analysis of clinical symptoms and signs of arsenism caused by coal burning pollution (in Chinese with English abstract). Chin J Endemiol 19(2):139–141.

Linak, WP, Wendt-Jost, OL (1994) Trace metal transformation mechanisms during coal combustion. Fuel Process Technol 39:173–198.

Liu, D, Yan, Q, Tang, D (2001) Geochemistry of sulfur and elements in coal from the Antaibao surface coal mine, Pingshuo, Shanxi Province, China. Int J Coal Geol 36(1):51–64.

Liu, GJ (2000) Experiment of coal leaching and study of the separation of trace elements. Acta Geol Sin 74(2):386–390.

Liu, GJ, Yang, PY (1999) Geochemistry of trace elements in Jining coalfield (in Chinese with English abstract). Geol Geochem 27(4):77–82.

Liu, GJ, Peng, ZC, Yang, PY, Wang, G (2001a) Sulfur in coal and its environmental impact from Yanzhou Ming District, China. Chin J Geochem 20(3):274–281.

Liu, GJ, Peng, ZC, Yang, PY, Gui, H, Wang, G (2001b) Characteristics of coal ash in Yanzhou Mining district and distribution of trace elements in them. Chin J Geochem 20(4):357–367.

Liu, GJ, Peng, ZC, Wang, G, Chou, CL (2002a) Study on the trace elements in coal (in Chinese with English abstract). Adv Earth Sci 17(1):53–62.

Liu, GJ, Yang, PY, Peng, ZC, et al. (2002b) Occurrence of trace elements in coal of Yanzhou mining district (in Chinese with English abstract). Geochimica 31(1):85–90.

Liu, GJ, Peng, ZC, Zhang, HY (2003) Characterization and leaching behavior of fly ash from power plants in Yanzhou district, China. Geochim Cosmochim Acta 67(18)(suppl):A257.
Liu, GJ, Yang, PY, Peng, ZC, Chou, CL (2004a) Petrographical and geochemical contrasts and environmentally significant trace elements in marine-influenced coal seams, Yanzhou Mining Area, China. J Asian Earth Sci 23(4):491–506.
Liu, GJ, Yang, PY, Peng, ZC (2004b) Petrological and mineralogical characterizations and chemical composition of coal ashes from power plants in Yanzhou Mining District, China. Fuel Process Technol 85(15):1635–1646.
Liu, GJ, Zhang, HY, Yang, PY, Peng, ZC (2004c) Volatility of trace elements and characterization of ash at different coal combustion temperatures, Proc 3rd Int Symp Heat Transfer Energy Conserv 1:375–380.
Liu, GJ, Zheng, LG, Peng, ZC (2004d) Distribution of hazardous trace elements during coal beneficiation. Geochim Cosmochim Acta 67(18)(suppl):A522.
Liu, J, Wu, X, Yao, Q, et al. (1999a) Study on emission and retention of fluorine during coal combustion in the chain-grate furnace (in Chinese with English abstract). J Eng Thermophys 20(5):642–646.
Liu, J, Yao, Q, Cao, X, et al. (1999b) Research on the measurement and regularities of distribution of fluorides in coal (in Chinese with English abstract). Coal Geol Explor 27(2):9–12.
Liu, J, Qi, Q, Cao, X, et al. (2000) Studies on existence pattern, combustion transfer and retention mechanism of fluorine in coal. China Sci Technol Period 6(10):319–321.
Lu, B (1996) Modes of occurrence of fluorine and chlorine in China coal (in Chinese with English abstract). Coal Geol Explor 24(1):9–11.
Lu, X, Zeng, H, Ouyang, Z (1995) The distribution and enrichment of trace elements in particles released from coal combustion (in Chinese with English abstract). Environ Chem 14(6):489–493.
Lu, X, Ali, A, Liu, H (1996) Study on the speciation distribution of heavy metals in particulates released from coal combustion (in Chinese with English abstract). Environ Chem 15(4):337–342.
Luo, KL, Li, R, Wang, L (2001) Fluorine content and distribution law of coal and its source in China (in Chinese with English abstract). Environ Sci (suppl):26–31.
Luo, KL, Xu, L, Li, R (2002) Fluorine emission from combustion of steam coal of north China plate and northwest China. Chinese Sci Bull 47(16):1346–1350.
Luo, KL, Ren, D, Xu, L, Dai, S, Cao, D, Feng, F, Tan, J (2004) Fluorine content and distribution pattern in Chinese coals. Int J Coal Geol 57:143–249.
Mastral, AM, Callen, MS, Garcia, T (2000) Toxic organic emissions from coal combustion. Fuel Process Technol 67:1–10.
Nerin, C, Zufisurre, R, Cscho, I (1994) Arsenic as long distance tracer of air pollution. Toxicol Environ Chem 44(1–2):33–42.
Ni, B (2000) New progress in high-precision and high resolution seismic exploration technology in coal industry of China. Acta Geol Sinica 74(2):311.
Notcutt, G, Davies, F (2001) Environmental accumulation of airborne fluorides in Romania. Environ Geochem Health 23(1):43–51.
Nriagu, JQ, Pacyna, JM (1988) Quantitative assessment of worldwide contamination of the air, water and soils with trace metals. Nature (Lond) 333:134–139.

Piekos, R, Paslawska, S (1999) Fluoride uptake characteristics of fly ash. Fluoride 32(1):14–19.
Querol, X, Fernandez-Turiel, JL, Lopez-Soler, A (1995) Trace elements in coal and their behavior during combustion in a large power station. Fuel 74:331–338.
Ren, D, Fenghua, Z, Wang, Y, et al. (1999) Distributions of minor and trace elements in Chinese coals. Int J Coal Geol 40:109–118.
Rizeq, RG, Hansell, DW, Beeker, WR (1994) Predictions of metals emission and partitioning in coal fired combustion systems. Fuel Process Technol 39: 219–236.
Sun, B, Nie, G, Zhang, Y, et al. (1984) Investigation report of chronic arsenism: material collection of sanitation and epidemic prevention (in Chinese with English abstract). Sanitation and Antiepidemic Station of Guizhou Province, pp 284–289.
Swaine, DJ (1990) Trance Elements in Coal. Butterworths, London, pp 109–386.
Swaine, DJ (1994) Trace elements in coal and their dispersal during combustion. Fuel Process Technol 39:121–138.
Swaine, DJ (2000) Why trace elements are important. Fuel Process Technol 65–66:21–23.
Swaine, DJ, Goodarzi, F (1995) Environmental Aspects of Trace Elements in Coal. Kluwer, Dordrecht.
Su, H, Yan, L, Rao, S, Jian, X, Mao, D (1990) Investigation of the cause of the origination of the environmental selenium area in the Exi Autonomous Prefecture of Hubei Province. Environ Sci 11(2):86–89.
Sun, J, Jervis, RE (1986) Distribution characterization of trace elements during coal combustion. China Sci Bull (A) 16:1285–1294.
Tang, X, Huang, W (2002) Trace elements of coal and its significances on research. Coal Geol China 14(suppl):1–4.
Tang, W, Gu, G, Zeng, X, et al. (1999a) Study on fluorine migration and transformation and its control countermeasure to power plant burning coal (in Chinese with English abstract). Environ Protect 2:13–15.
Tang, W, Gu, G, Zhao, J, et al. (1999b) Study on fluorine pollution problem of power plant burning coal (in Chinese with English abstract). Chongqing Environ Sci 21(4):39–42.
U.S. National Committee for Geochemistry (1990) Panel on the Trace Elements Geochemistry of Coal Resource Development Related to Health, Trace Element Geochemistry of Coal Resource Development Related to Environmental Quality and Health. National Academy Press, Washington, DC, pp 10–68.
Wang, H, Hao, C, Wang, J (1992) Asenic in environment (in Chinese). Press of Environment Science of China Beijing, pp 30–88.
Wang, Q, Kang, S, Chen, C, Wang, Z, Zou, S (1996) Study on the contents and distribution laws of trace elements in coal in Northeast China and Eastern Inner Mongolia (in Chinese with English abstract). Environ Chem 15(1):27–35.
World Bank (1992) The World Bank Development Report 1992. Development and Environment, Washington, DC.
Xu, X-C, Chen, C, Qi, H (2000) Development of coal combustion pollution control for SO_2 and NOx in China. Fuel Process Technol 62:153–160.

Xu, Y, Zhai, J (1997) Distribution and enrichment of trace elements in particles of power fly ash with different size from Huaneng Nanjing Poer Plant (in Chinese with English abstract). Geochimica 26(3):73–77.
Yan, R, Gautheir, D, Flamant, G, et al. (1999) Thermodynamic study of the behavior of minor coal elements and their affinities to sulfur during coal combustion. Fuel 78:1817–1829.
Yang, G, Zhou, R, Sun, S, Yin, T, Liu, S (1981) Study of unidentified trichomadesis and nail losing of Enshi Prefecture of Hubei Province. J Acad Med Sci China 3(suppl):1.
Yang, S, Qian, Q (1983) The distribution of elements in ashes during coal combustion in power plants (in Chinese with English abstract). Environ Chem 2(2):32–37.
Zhang, J, Wang, Y, Zhang, R, et al. (1999) Distribution of arsenic in coal and its resides (in Chinese with English abstract). Res Environ Sci 12(1):27–30.
Zhang, JY, Zheng, CG, Ren, DY (2004) Distribution of potentially hazardous trace elements in coals from Shanxi Province, China. Fuel 83:129–135.
Zhang, Y, Cao, SR (1996) Coal burning induced endemic fluorosis in China. Fluoride 29(4):207–211.
Zhao, L, Zhang, B (1988) Geochemistry. Geological Publishing House, Beijing, pp 60–81.
Zhao, F, Ren, D, Zheng, B (1998) Modes of occurrence of arsenic in high arsenic coals from EXAFS spectroscopy (in Chinese with English abstract). Sci Bull 43(14):21–23.
Zhao, F, Ren, D, Peng, S (2003) The mode of occurrence of arsenic in coal (in Chinese with English abstract). Adv Earth Sci 18(2):214–220.
Zheng, B (1992) Study of Endemic Fluorosis and Industrial Fluorine Pollution (in Chinese with English abstract). Chinese Environmental Science Press, Beijing.
Zheng, B, Cai, R (1988) Study on fluorine content in China coal (in Chinese). Chin J Control Endemic Dis 3(2):70–72.
Zheng, B, Huang, R (1984) Analysis of disease cause of fluorosis in human beings and livestock in parts of China In: First Annual Meeting of the Association of Environmental Science of China, Beijing, abstract, p 13 (in Chinese with English abstract).
Zheng, B, Huang, R (1985) Environmental geochemistry study of the fluorosis caused by burning coal indoors in southwestern China (in Chinese with English abstract). News Dispatch of Endemic Diseases, vol 3, pp 49–51.
Zheng, B, Huang, R (1986) Prevention and cure and study of the fluorosis caused by pollution of domestic coal (in Chinese with English abstract). J Pract Endemic Dis Study 1(2):11–13.
Zheng, B, Huang, R (1989) Human fluorosis and environmental geochemistry in southwest China (in Chinese). Developments in Geoscience. Contributions to 28th International Geologic Congress, Washington, DC. Science Press, Beijing, China, pp 171–176.
Zheng, B, Zhang, J, Zhou, D (1996) Environmental geochemistry of coal and endemic arsenism in southwest Guizhou, P.R. China. 30th Int Geol Congr Abstr 3:410.
Zheng, B, Ding, Z, Huang, R, et al. (1999) Issues of health and disease relating to coal use in southwestern China. Int J Coal Geol 40:119–132.

Zhong, T, Yang, W (2000) Efficient and environment friendly use of coal. Fuel Process Technol 62:137–141.

Zhou, DX, Liu, DN, Zhu, S (1993) Investigation of chronic arsenism caused by pollution of high arsenic coal (in Chinese with English abstract). J Chin Prevent Med 27(3):147.

Zhou, Y (1998) Occurrence and distribution of Asenic in coal from old mines (in Chinese with English abstract). Coal Geol Explor 26(4):8–13.

Zhou, Y, Du, H, Cheng, M, et al. (2002) The investigation of death disease caused by coal burning type arsenic poisoning (in Chinese with English abstract). Chin J Endemiol 21(6):484–486.

Manuscript received March 4; accepted March 16 2006.

Rev Environ Contam Toxicol 189:107–130

Mercury Content of Hair in Different Populations Relative to Fish Consumption

Krystyna Srogi

Contents

I. Introduction 107
II. Biomarkers of Human Exposure to Pollutants 109
III. Hair Analysis as Biomarker of Exposure to Mercury 109
IV. Inorganic Mercury and Methylmercury: History 110
A. Inorganic Mercury 110
B. Methylmercury 111
V. Relationship Between Mercury Concentrations in Blood and Hair in Methylmercury Exposure Subjects 113
VI. Human Exposure Case Studies 113
A. Minamata Disease in Japan 113
B. Asian Populations 115
C. Amazonia 116
D. Sweden 117
E. Alaska 118
F. Arabia 119
G. Other 119
H. Organic Mercury Poisoning in Iraq 121
VII. Conclusions 121
Summary 121
References 122

I. Introduction

Mercury is ranked third behind lead and arsenic on the ATSDR/EPA priority list of hazardous substances (ATSDR 1997) based on toxicity and prevalence at contaminated sites. The hazardous effects of mercury initially became well known because of the major mercury-related accident that contaminated several humans in Minamata, Japan (Tsubaki and Irukayama 1977).

In the environment, elemental mercury may undergo oxidation to inorganic mercury IHg and biotransformation to methylmercury (MeHg). MeHg is a naturally occurring environmental compound that bioaccumulates in

---

Communicated by David M. Whitacre.

K. Srogi (✉)
Silesian University of Technology, Department of Inorganic Chemsitry, Krzywoustego 6, 44–100 Gliwice, Poland

fish and piscivorous species; MeHg comprises greater than 95% of total mercury in edible fish tissues (Stewart and Stolman 1961; Clarkson et al. 1988; Clarkson 1997; Mason et al. 1995; Bloom 1992). High exposure to MeHg during fetal development has been linked to adverse human development, including mental retardation and altered motor function (Matsumoto et al. 1965; Choi et al. 1978).

In humans, the main target for mercury, specifically MeHg, is the central nervous system, and exposure can cause serious brain damage (WHO 1990), including psychological disturbances, impaired hearing, loss of sight, ataxia, loss of motor control, and general debilitation (Blanyney 2001; Dixon and Jones 1994). Several cases have been lethal (Nierenberg et al. 1998). In addition, demethylation occurring in the organism has an influence on the ratio of inorganic to organic mercury compounds in various tissues (Mempel et al. 1995; Watt 2001; Philip and Oyuah 2000). Lind et al. (1988), in an experiment performed on monkeys that were exposed for several years to methylmercury, found at the end of the exposure period that 10%–30% of the mercury in the brain was in an inorganic form, but 2 yrs after the exposure terminated, 90% was in an inorganic form. This result confirms the existence of demethylation processes in the brain.

Due to increasing fish consumption (Wheeler 1996) and the high capacity of mercury for bioaccumulation and biomagnification, fish are the most critical source of mercury for human beings. Because of its relevance to public health, biomarkers of MeHg exposure have been developed (Fiedler et al. 1999). These biomarkers include Hg content of blood and hair (Seppänen et al. 2000), which are specific and sensitive measures of exposure following prolonged MeHg exposure (Orloff et al. 1997; Grandjean et al. 1994; Phelps et al. 1980; Lind et al. 1988; Chang 1996).

Foo et al. (1993) reported that hair Pb, Hg, and manganese (Mn) content of blood and urine are useful biological indicators for exposure as they correlated with blood Pb, blood Hg, or urinary Mn. Element residues in hair also provide a lasting record (Suzuki and Yamamoto 1982) of exposure over the previous few months. They are particularly useful for assessing exposure months after the exposure has ceased.

The content of total mercury (THg) or MeHg in hair is usually well correlated with mercury blood levels. In an unexposed population when the daily intake of mercury did not exceed 10 mg mercury, the hair-to-blood ratio was between 200:1 and 300:1 (Moszczynski and Moszczynski 1990). The content of essential and toxic elements in hair has been determined by many authors. They concluded that the interpretation of results is difficult unless a correlation between the concentration in hair and a critical organ is ascertained. However, Muramatsu and Parr (1988), in autopsy material from 30 Swedish subjects, found a positive correlation between the levels of mercury and selenium in the hair and renal cortex, as well as a correlation between the mercury content in renal cortex and liver. Suzuki et al. (1993), in 46 autopsy samples from the inhabitants of Tokyo, determined a correlation between mercury content in hair and in the cerebrum, heart, spleen, liver, kidney cortex, and kidney medulla (Hać et al. 2000).

II. Biomarkers of Human Exposure to Pollutants

As environmental pollution increases, the risks to health from this pollution are of increasing concern to the public. It is the role of environmental toxicology to address the risks of environmental pollution to human health. Analysis for pollutants in suitable human tissue appears to be more appropriate than the analysis of plant (Srogi 2004) or animal material in defining potential risks. The human tissues most accessible for sampling include blood and urine (Muszyńska-Zimna 1982), and also hair (Srogi 2005) and nails (Nowak 1996), the derivatives of the ectoderm (Bencko 1995).

III. Hair Analysis as a Biomarker of Exposure to Mercury

Human hair grows at a rate of approximately 1 cm/mon (Cernichiari et al. 1995). The element level in hair reflects in the body and provides a historical record of elements assimilated from the environment over time (Buzina et al. 1995; Wiadrowska et al. 1983; Wiadrowska and Syrowatka 1983). Hair samples are easy to collect, transport, and store (Airey 1983a), and the metal content in hair has been reported to be stable for long periods of time (Phelps et al. 1980; Suzuki and Yamamoto 1982). In general, hair also contains a higher concentration of metals than blood or urine, which makes analysis easier (Airey 1983b). Suzuki (1988) listed noninvasive sample taking, stable during storage, higher concentration, and long term record of exposure as advantages for using hair to estimate the absorption of metals. The body burden of toxic elements in members of the community is mainly derived from exposure to the environment (community air, water, and dietary products). Occupational exposure may also contribute to the body burden, but unless there are special circumstances, e.g., living near a lead smelter, occupational exposure is not the major contributor. Absorbed toxic elements not only affect the person directly exposed but may also affect developing fetuses through cross-placental transportation (Huel et al. 1984; Ong et al. 1993).

Environmental mercury vapor may bind to hair (Yamaguchi et al. 1975), and permanents from hair dressers may remove endogenous mercury from the hair (Yamamoto and Suzuki 1978; Yasutake et al. 2003). Also, hair color may affect the mercury concentration (Grandjean et al. 2002).

MeHg is incorporated into hair during its growth in the follicle, but requires at least a month to become detectable in a hair sample cut with scissors close to the scalp (Grandjean 1984; Suzuki 1988). Depending on the length of the hair sample, mercury concentration may reflect exposures that extend far into the past. Still, because the half-life of methylmercury in the body is about 1.5–2 months (Smith and Farris 1996; Swartout and Rice 2000), the hair closest to the scalp reflects recent exposures that also reflect to the current blood concentration (Grandjean et al. 2002). The association between the two exposure biomarkers (blood and hair) depends on the circumstances of individual studies, and, as a result, some prospective cohort studies on developmental methylmercury toxicity have used both biomarkers (Grandjean et al. 1999).

IV. Inorganic Mercury and Methylmercury: History

Mercury is both the best known and most puzzling of metals in the environment. It occurs in all media in several species, both organic and inorganic. It is relatively uncommon in the Earth's crust, from which it is liberated by natural processes such as erosion and vulcanism as well as by mining. Anthropogenic activities both intentional (mercury manufacture and disposal) and unintentional (fossil fuel combustion) contribute most of the mercury in the environment. The mercury species differ greatly in properties, but all are toxic. Elemental mercury (Hg^0) is a familiar dense, silvery liquid at room temperature. It volatilizes readily, emitting mercury vapor, which is readily absorbed in the lungs; however, elemental mercury is very poorly absorbed through the skin or gastrointestinal (GI) tract. Inorganic mercurial salts (Hg^+ and Hg^{2+}) vary in solubility and absorptive properties. Most organic mercurial compounds are readily absorbed through lungs and GI tract, and some are readily absorbed through the skin. All mercury compounds are toxic to humans and animals, but the organic forms, particularly methylmercury and dimethylmercury, are the most toxic. MeHg is the form found most widely in nature, and it bioaccumulates in the food chain. It is the form responsible for most human exposure (Gochfeld 2003).

A. Inorganic Mercury

Exposure to elemental mercury occurs in selected workplaces, in healthcare facilities, and occasionally in homes. Droplets of liquid mercury are attractive to humans, and children have been known to bring mercury home to play with. Cultural practices such as Santeria also result in household exposures to elemental mercury. Breakage of thermometers and spills from gas meters during their removal are infrequent but important sources of mercury. When such spills occur it is important that they be cleaned up quickly, because droplets of mercury continue to emit vapor for a long time. Exposure to elemental Hg is almost exclusively through inhalation of vapor.

Hg^0 vapor in the atmosphere is subject to long-range transport. Hg^0 is slightly soluble in water (0.08 mg/L at 25°C) (ATSDR 2000). Therefore a small fraction of Hg^0 vapor can be washed out of the atmosphere during precipitation. The more likely fate of Hg^0, however, is eventual oxidation to

Hg^{2+} by reaction with atmospheric oxidants such as oxygen, ozone, and chlorine. Once converted to the Hg^{2+} form, the Hg is much more soluble and more subject to washout from the atmosphere with precipitation; this is called wet deposition and is a major source of mercury input to the environment. A small amount of the mercury may adhere to fine particles in the atmosphere and may fall out without rainfall as dry deposition (Fitzgerald 1995). Today, the major sources of mercury for the general environment include the incineration of wastes and the burning of coal to produce electricity. Much of this inorganic mercury reaches aquatic systems (either as direct fallout to water surfaces or as indirect fallout to land with subsequent runoff). In bodies of water it settles onto the sediment surface or algal mats, where it is converted by bacterial action from inorganic forms into the highly toxic methylmercury (Gochfeld 2003).

B. Methylmercury

Elemental and inorganic mercury are deposited on watersheds or directly on bodies of water. This mercury settles to the sediment, where some of it reacts with sulfate to form an insoluble mercuric sulfide precipitate, whereas a small percentage is biomethylated by bacteria. This MeHg is readily bioavailable and biomagnifies up the food chain so that fish at higher trophic levels regularly have mercury concentrations a million fold greater than the water in which they live. Fish consumption poses the only significant source of dietary exposure to MeHg for most people, and conversely 75%–95% of the mercury in fish is in the MeHg form. Methylmercury exposure over a period of years resulted in the outbreak known as Minamata disease (Section VI.A) and continues to be a source of mercury poisoning among fish-eating peoples of Iraq (Section VI.H). People who eat large amounts of fish (even species with relatively low mercury content) can accumulate sufficient levels of MeHg to cause symptoms, and pregnant women can transfer to a fetus amounts of MeHg that are sufficient to impair nervous system development.

It should be concluded that, with relatively few exceptions, the general public has very low exposure to elemental mercury, and these exposures are often related to deliberate uses (e.g., occupational, dental, and perhaps folk/cultural practices). The health consequences of widespread mercury exposure from dental amalgams (Hörsted-Bindslev 2004) are currently highly controversial. Nylander et al. (1987) found in six Swedish amalgam carriers that 77% of the mercury in the brain was in an inorganic form. For most people, the greatest problem is with MeHg, to which the most significant exposure comes from eating fish. There is a large variation in the amount of fish that people consume as well as in the concentrations of mercury in different kinds of fish (Stern et al. 2001). The major sources and estimates of human exposures to various forms of mercury are given in Table 1 (Gochfeld 2003).

Table 1. Sources and estimates of daily human exposures to various forms of mercury (Gochfeld 2003).[a]

Source of exposure	MeHg (μg/d)	Inorganic Hg salts (Hg^{2+}) (μg/d)	Elemental Hg (Hg^0) (μg/d)
Food (non-fish)	Negligible	0.9	Negligible
Commercial fish	6	<1	Negligible
Sport fish	No population-based data available	No population-based data available	Negligible
Drinking water	Negligible	Most public supplies <<4 μg, but some wells >4 μg	Negligible
Outdoor air	Negligible	Negligible	0.2–0.4 μg
Indoor air	Negligible	Negligible	Cultural, recreational, and accidental uses in a few homes
Soil ingestion	Negligible	Contaminated sites may >3 μg	Negligible
Dental amalgams	Negligible	Negligible	<<1 μg but occupational exposure

Negligible is defined as 0.1 mg/d.
[a]Modified from New Jersey Mercury Task Force (2001).

V. Relationship Between Mercury Concentrations in Blood and Hair in Methylmercury Exposure Subjects

From a toxicological point of view, the blood concentration is the best indicator of an absorbed dose and the amount systemically available. Methylmercury binds to hemoglobin, and its high affinity to fetal hemoglobin results in a higher mercury concentration in cord blood than in maternal blood (Tsuchiya et al. 1984; Stern and Smith 2003). Although routine analyses of total mercury concentrations will include inorganic mercury, cord blood mercury is almost entirely in the methylated form, which passes easily through the placenta (Kelman et al. 1982; Tsuchiya et al. 1984; WHO 1990). In comparing results of these two biomarkers, temporal factors must also be considered.

Factor analysis has been used to determine the total error, i.e., laboratory imprecision and all preanalytical variation, of these two biomarkers (Budtz-Jørgensen et al. 2003). The total error of hair mercury determinations was found to be almost twice that of blood. This finding is in accordance with the toxicological prediction and empirical observation that the cord-blood mercury concentration is a better predictor of mercury-associated neurobehavioral deficits (Grandjean et al. 1999). However, even the cord-blood parameter had a surprising imprecision (expressed by the coefficient of variation, CV) of about 30%. A nondifferential error in the biomarker is likely to cause underestimation of the true association with the outcome variables (Carroll 1998). Adjustment can be made only if the extent of the total error is known. Translation between studies that have applied different biomarkers requires information about conversion factors. Under reasonably constant conditions of exposure, the human hair-mercury concentrations (in μg/g) are thought to average about 250 fold the whole blood mercury concentrations (in μg/mL) (US EPA 2001). However, concentration ratios expressed as regression coefficients appear to vary somewhat in European and American population groups (Phelps et al. 1980; Turner et al. 1980; Akagi et al. 1998). A limitation of regression analyses is that they do not take into account that both biomarkers are affected by imprecision, and the concentration ratio will therefore depend on the choice of the dependent variable (Ponce et al. 1998).

VI. Human Exposure Case Studies

A. Minamata Disease in Japan

MeHg poisoning was first recognized in Minamata, Japan around 1960. Hundreds of fishermen and their families were severely poisoned during the 1950s by MeHg that bioaccumulated in fish as a result of release of mercury to the bay from a local chemical plant. By 1956 the affected population recognized that their poisoning was due to fish and abandoned

Table 2. Levels of mercury (Hg) in human hair from different populations.

Location	Levels of Hg compounds (μg/g)		Comments	Reference
Papua New Guinea	MeHg	15.5 (mean)	Heavy fish–consuming population	Kyle and Ghani 1983
		3.2–50.5 (range)		
	MeHg	6.4 (mean)	Lesser fish–consuming population	
		0.62–25.7 (range)		
	MeHg	2.4 (mean)	Control group	
		0.33–9.0 (range)		
Papua New Guinea	THg	21.9 (mean)	Fish-eating population	Abe et al. 1995
		3.7–71.9 (range)		
Finland	THg	4.9 (mean)	Middle-aged fish-eating population	Lodenius et al. 1983
Spain	THg	10.41 and 8.36 (mean)	Fishermen	Lopez-Artiguez et al. 1994
Spain	THg	0.77 (mean)	Children (6–16 yr)	Batista et al. 1996
		0.18 ± 2.44 (range)		
Kuwait	THg	4.60 ± 4.80 (mean)	Kuwait residence age (2–57 yr)	Bou-Olayan and Al-Yakoob 1994
		0.80 ± 25.00 (range)		
Columbia	THg	5.23 ± 5.78 (mean)	Fishermen	Olivero et al. 1995
		2.38 ± 3.27 (mean)	Miners	
		2.40 ± 2.02 (mean)	People of various activities	
		1.33 ± 0.78 (mean)	Control samples	
Japan	THg	2.08–36.5 (range)	Diseased volunteers	Ryozo 1995
Florida	THg	3.62 ± 3.00 (mean)	Consumption of fish rate	Fleming et al. 1995
		1.28–15.57 (range)	One meal/wk	
Italy	THg	>6.0	Fishermen	Valentino et al. 1995
Peru	THg	8.3 (mean)	Infant–mother pairs	Marsh et al. 1995
		1.20–30.0 (range)	Fish-eating population	
Seychelles	THg	5.9–8.2 (range)	Fish-eating population	Cernichiari et al. 1995
Croatia	THg	1.3–12.9 (range)	Age (2–83 yr)	Buzina et al. 1995
	MeHg	1.1–10.8 (range)		
Brazil	THg	4.7–151 (range)	General population	Malm et al. 1995
Bangladesh	THg	0.44 ± 0.19 (mean)	Age (13–69 yr)	Holsbeek et al. 1996
Denmark	THg	4.5 (mean)	Fish-eating population	Weihe et al. 1996
Egypt	THg	0.23 ± 0.06	Group without dental amalgam	Mortada et al. 2002
Hong Kong	THg	0.38 (mean)	All population (vegetarian)	Dickman et al. 1998
Chinese (Singapore)	THg	4.38 (mean)	All population	Foo et al. 1988
Indonesians (Batam)	THg	2.72 (mean)		

MeHg, methyl mercury; THg, total mercury.

eating fish from the bay. However, cases continued to accumulate, and by 1960 the serious and mysterious affliction, affecting both adults and infants, was recognized as MeHg poisoning, a hitherto unrecognized disease. The Minamata epidemic, first reported by Kurland et al. in 1960, has been studied more recently by Eto (2000), Futatsuka et al. (2000), Kondo (2000), Harada (1995), Harada et al. (1998), and Akagi et al. (1998).

High-level exposure produced serious neurological disease in adults, but the most dramatic manifestation was congenital Minamata disease in infants born to mothers with high mercury levels. These babies were born with severe cerebral palsy, blindness, and profound mental retardation. Many of these victims survived into their twenties, and many of the adult victims survive to this day. Autopsies of victims revealed destructive lesions of the cerebellum and in various parts of the cortex. Peripheral neuropathy also occurred. A selective summary of Hg in hair data is given in Table 2.

B. Asian Populations

Large differences have been reported among Asian workers for hair concentrations of Hg (0.44–21.9 μg/g). The differences in hair Hg are likely to be caused by the diversity of ecological factors, community, family, and individual practices, and dietary intake. Hg hair content for the Indonesians in Batam was lower than that of the Singapore Chinese or Malays. The Singapore Malays and Batam Indonesians are closely related ethnically. Moreover, Singapore and Batam Island are geographically close (approximately 15 km apart) and were one political entity before British rule in Singapore. The difference in hair Hg levels might reflect the different dietary and environmental pollution patterns as a result of the different stages of economic development between the islands. The reasons for the lower hair Hg levels in Batam islanders were not fully understood. No clear geographic pattern for hair Hg levels can be observed from reported data in Asian countries (Takeuchi et al. 1982; Ohno et al. 1984; Matsubara and Machida 1985; Sarmani et al. 1985; Foo et al. 1988; Srikumar et al. 1992; Bou-Olayan and Al-Yakoob 1994; Abe et al. 1995; Foo and Tan 1998). The results of the study were consistent with data published for Singapore and Indonesia (Ohno et al. 1984; Foo et al. 1988). Even though the dietary fish consumption pattern was reported to have an effect on Hg levels in hair (Foo et al. 1988; Abe et al. 1995), these differences cannot be explained by fish consumption alone as the residents in islands off Batam are mainly fishermen and fish is part of the regular diet (Foo and Tan 1998). In conclusion, the environmental exposure to Hg was higher for the Singapore residents compared to the fishing communities on the Batam isles. However, the Batam fishing communities were exposed to higher levels of lead and cadmium compared to Singapore residents. Lead exposure at the fishing communities were likely related to the making of lead sinkers for the fishing nets.

Nakagawa (1995) concluded that the mean THg, 2.23 μg/g, of volunteers from Tokyo and the surrounding areas was considerably lower when compared to those living in the same areas in the 1960s, 6.02 μg/g. This difference may be attributed to the change in eating habits, the decrease in industrial usage for Hg, and the reduction in unlawful discharge of Hg wastes. Contemporary Japanese youth tend to prefer Western foods to traditional Japanese foods, which are mainly composed of fish.

C. Amazonia

There has been increasing evidence of Hg pollution in the Amazon Basin and high hair Hg levels among riverine communities, for whom fish constitute the dietary mainstay. In the Amazon Basin, Hg contamination of fish has been recognized as a problem affecting many of the indigenous people who live along the tributaries and rely on fish as a major source of protein. Hg has been released during gold mining operations, but the largest sources may be deforestation and damming of rivers, which makes naturally occurring Hg in the soil accessible to biomethylation by aquatic organisms (Roulet et al. 1999; Barbosa and Dorea 1998).

Studies from different areas in the Brazilian Amazon indicate that hair mercury median values vary between 2 μg/g and 20 μg/g (Akagi et al. 1995; Barbosa et al. 1995 1998; Malm et al. 1995; Boischio and Henshel 1996; Boischio and Cernichiari 1998) or between 8.7 μg/g and 75.5 μg/g (Barbosa and Dorea 1998). Studies on populations living along the Madeira River reported that Hg content in human hair reached levels as high as 27 μg/g (Malm et al. 1990).

Recent studies have shown that neurotoxic effects are associated with long-term exposure to mercury at low levels (Dolbec et al. 2000). Dolbec et al. (2001) were able to document that hair Hg levels among Amerindian women varied with the season. During the wet season when herbivorous fish predominated in the diet, levels were lower than in the dry season, when herbivores made up <50% of the diet. They also found a positive association with reported fish consumption; however, paradoxically, the proportion of hair Hg that was inorganic increased with increasing fish consumption. They also reported cytogenetic damage in lymphocytes in women with hair Hg <50 μg/g. Guimaraes et al. (2000) demonstrated spatial and seasonal variation in biomethylation of mercury using an isotope method. Using ^{203}Hg as a tracer, they showed that MeHg was formed in the superficial layer of flooded soil or sediment and on the surface of aquatic plants (macrophytes). In a cross-sectional study, Dolbec et al. (2000) studied neurobehavioral performance of 84 residents along the Rio Tapajos (Bidone et al. 1997), using hair Hg as their exposure metric. The mean percentage of meals containing fish was 62%, and the median hair Hg was 9 μg/g (~95% was organic Hg). There was a negative correlation between hair Hg and neurobehavioral performance, particularly in fine motor coordination, and the relationship was stronger than with blood Hg, indicating the benefits of using hair levels that track exposure over a period of months. However, only 8%–16% of the variance in performance was explained by the hair Hg. Cytogenetic damage also correlated with hair Hg levels (Amorim et al. 2000).

Santos et al. (2002) have reported a similar study on a Brazil population, Munduruku Indians from the community of Sai Cinza, State of Pará. The population of Munduruku Indians showed high concentrations of total Hg in hair, ranging from 4.50 up to 90.40 μg/g. The presented data (Santos et al. 2002) was observed for all persons with no statistical differences by age group, sex, occupation, or diet. There was a weak and positive correlation in hair Hg among women of fertile age and their children. The area in which this group lives is at risk of pollution by Hg from gold extraction. These high levels could be increased by the cultural practices of the Mundurukus. Fish is the main food consumed, with continuous fish ingestion throughout the day for all ages. This pattern of exposure is similar to that of other riverside (non-Indian) communities from the Tapajós River basin, such as Saão Luiz do Tapajós, which is at risk of pollution by Hg from gold mining (Santos et al. 2000). It is necessary to know the importance of other variables that could be related to the mechanisms of MeHg absorption by humans, such as the possible influence of alimentary habits. Also important to the health conditions of this population is the high prevalence of intestinal parasitism in addition to endemics such as malaria that weaken individuals, organisms affecting their immunological system and possibly their exposure to Hg. The concentrations of total Hg in fish in Sai Cinza were below the limit recommended by the World Health Organization (WHO).

D. Sweden

Ask Björnberg et al. (2005) have reported studies of fish consumption and Hg exposure. They studied exposure to MeHg in 127 Swedish women with high fish consumption (eat fish several times/mon), 19–45 yrs of age (median, 38 yrs), using THg in hair and MeHg in blood as biomarkers. The average total fish consumption, as reported in the food frequency questionnaire (FFQ), was approximately 4 times/wk (range, 1.6–19 times/wk). Fish species potentially high in MeHg, included in the Swedish dietary advisories, was consumed by 79% of the women. About 10% consumed such species more than once a week, i.e., more than recommended. Other fish species potentially high in MeHg, not included in the Swedish dietary advisories, were consumed by 54% of the women. Eleven percent never consumed fish species potentially high in MeHg. THg in hair (median, 0.70 mg/kg; range, 0.08–6.6 mg/kg) was associated with MeHg in blood (median, 1.7 μg/L; range, 0.30–14 μg/L). Hair THg, blood MeHg (median, 70 μg/L; range, 46–154 μg/L) increased with increasing total fish consumption. IHg in blood (median, 0.24 μg/L; range, 0.01–1.6 μg/L) increased with increasing number of dental amalgam fillings. They found no statistical significant associations between the various mercury species measured. Hair Hg levels exceeded the levels corresponding to the EPA reference dose of 0.1 μg MeHg/kg b.w./d in 20% of the women. Thus, there seems to be no margin of safety for neurodevelopmental effects in the fetus for women with high fish consumption unless they decrease their intake of certain fish species.

Lindberg et al. (2004) have investigated the THg exposure of Sweden population (Stockholm) who do not consume fish or fish products and found that the THg concentration in hair (median, 0.06 mg/kg; range, 0.04–0.32 mg/kg) was significantly lower ($P < 0.001$) than that in their previous study of pregnant women with a relatively low fish consumption (median, 0.35 mg/kg; range, 0.07–1.5 mg/kg; Ask Björnberg et al. 2003). Despite the low fish consumption of the pregnant women, especially of predatory fish species known to be potentially high in MeHg, possibly due to the recommendations for restricted fish intake for pregnant women given by the Swedish National Food Administration (SNFA 2002), the concentration in this study was about 15% of the prior levels. It has been assumed that THg in hair reflects IHg exposure in low- or non-fish-consuming populations (NRC 2000). However, because they found no correlation between the THg levels in hair and the IHg in blood and a relatively strong correlation between the THg in hair and the MeHg in blood, they concluded that THg levels in hair reflect MeHg and not IHg exposure also in individuals with no fish consumption and very low MeHg exposure.

E. Alaska

In Alaska, concentrations of Hg in freshwater are well below reported effects levels; however, the general relationship between effects and diet-related Hg exposure to humans are still unclear (Egeland and Middaugh 1997). A survey in 1976 showed that Hg concentrations in fishes and marine mammals consumed by Eskimos of the Yukon–Kuskokwim Delta were higher than in urban Eskimos of Anchorage, Alaska (Galster 1976). Over the past 15 yrs several studies have focused on the Hg content of water, sediment, and fishe in western Alaska, particularly from National Wildlife Refuges (Mueller et al. 1993, 1996; Mueller and Matz 2002). Among the various fishe analyzed for Hg, the greatest database exists on the freshwater fishe, northern pike (*Esox lucius*), and Arctic grayling (*Thymallus arcticus*). Tissues of the piscivorous northern pike have sporadically had THg concentrations that exceeded the U.S. Food and Drug Administration (FDA) action level for human consumption (1 mg/kg w.wt.). For example, 44% of the pike examined from the Nowitna National Wildlife Refuge in 1987 had concentrations in tissues between 1.0 and 2.9 mg/kg w.wt.

Duffy et al. (2001) concluded that, in Alaska, hair analysis (reindeer from the Seward Peninsula) of captive caribou ($n = 2$, 3.6 ng/g) and the free-ranging caribou indicated that free-ranging northcentral caribou were more exposed to Hg ($n = 11$, 47.1 ng/g; Duffy et al. 2001). The results (55.3 ng/g, $n = 5$) obtained by Duffy et al. (2005) indicate that free-ranging western Alaska reindeer have similar THg in their hair. Duffy et al. (2001) showed that the free-ranging caribou from the Teskekpuk Lake region of north-central Alaska showed some yearly variation in mean levels [1997, 47.1 ng/g ($n = 11$); 1998, 62.7 ng/g ($n = 11$)].

F. Arabia

A number of workers have investigated the importance of fish as a source of exposure to Hg in the Arabian (Persian) Gulf region. For example, in an extensive International Atomic Energy Agency exercise during 1984–1985 (ROPME 1988), total Hg values in fish of the region were estimated to range between 0.007 µg/g (w.wt.) for Mediterranean amberjack (*Seriola dumerlii*) and 0.540 µg/g (w.wt.) in Karanteen sea bream (*Crenidens crenidens*) (ROPME 1988). The overall mean value for different species was 0.158 ± 0.145 µg/g (w.wt.). Human exposure to Hg vapor in Kuwait was first studied by Zaki and Abdulraheem (1973) at the chloralkali plant that was in full operation at that time. However, these workers did not record data on blood levels or human hair content of Hg. Among Kuwait residents, Bou-Olayan and Al-Yakoob (1994) reported mean Hg concentrations in hair of 4.05.4.40 µg/g in 68 females and 5.52 ± 5.33 µg/g in 38 males. Samples represented individuals whose ages ranged from 2 to 57 yrs, with 21% above 6.0 µg/g.

Al-Majed and Preston (2000) have reported that the levels of THg and MeHg in the hair of about 78% of the fishermen and 63% of the control population exceeded 2.0 µg/g. In general, the mean concentrations of THg and MeHg are around twice the WHO normal level (2.0 mg µg/g) but are still less than the WHO threshold level (10.0 µg/g). However, it should be noted that levels were for fishermen who eat at least one meal of fresh fish per day, which is not the case for the general population. It may be added that the health risk of eating fresh fish may be less than that associated with eating canned tunafish, as may be concluded by comparing those eating two meals/wk of canned tuna fish (4.304 ± 0.708 and 4.224 ± 0.696 µg/g for total and MeHg, respectively) with those who eat seven meals of fresh fish/wk (3.658 ± 2.271 and 3.522 ± 2.210 µg/g for total and MeHg, respectively). In accordance with the strong correlation found between THg and MeHg ($r = 0.999$), one of these determinations can be used for human hair samples to determine the body content and, preferably, the MeHg because of its higher toxicity (WHO 1990). THg and MeHg concentrations in hair samples are found to be unrelated to age and the duration of residence in Kuwait for the Egyptian fishermen and the control group.

G. Other

Batista et al. (1996) have shown elevated hair Hg levels in populations from Spain living in different areas of Tarragona Province (Flix, Tarragona, and

Tortosa). They concluded that observed differences was probably due to the different food consumption by those populations. Although fish intake was similar for the three communities, people from Flix consumed higher amounts of frozen fish and seafood than the inhabitants from Tarragona and Tortosa. The average Hg levels in frozen fish consumed in Span (55 μg/kg w.wt.) (Moreiras and Cuadrado 1992) are remarkably lower than those from fresh fish captured along the Tarragona coast, 219 μg/kg w.wt. (Schuhmacher et al. 1994). Moreover, although the Hg levels in air, water, and vegetables from Flix were higher than those found in the other two communities examined, most Hg would be in the form of IHg. Because MeHg has higher retention (90% absorption) compared with IHg (10% absorption), the hair Hg levels depend mainly on the levels found in fish, as well as the frequency of fish and seafood consumption.

Holsbeek et al. (1996) reported that the Bangladesh hair Hg concentrations were very low, despite the fact that fish consumption was rather high. THg values ranged from 0.02 to 0.95 μg/g and fish consumption from 1.4 (farming community, Bogra) to 2.6 μg/g kg/mon (fishermen). A clear positive correlation was found between hair Hg levels and fish consumption patterns; in the absence of important natural or anthropogenic sources, fish consumption was found to be the only decisive factor for Hg intake.

Hg concentrations in human hair and fish samples from Phnom Penh, Kien Svay, Tomnup Rolork, and Batrong, Cambodia were also investigated by Agusa et al. (2005). Hg concentrations in human hair ranged from 0.54 to 190 mg/g dry wt. About 3% of the samples contained Hg levels exceeding the no observed adverse effects level (NOAEL) of WHO (50 mg/g), and the levels in some hair samples of women also exceeded the NOAEL (10 mg/g) associated with fetus neurotoxicity. A weak but significant positive correlation was observed between age and Hg levels in hair of residents. Hg concentrations in muscle of marine and freshwater fish from Cambodia ranged from <0.01 to 0.96 mg/g w.wt. Hg intake rates were estimated on the basis of the Hg content in fish and daily fish consumption. Three samples of marine fish including sharp-tooth snapper and obtuse barracuda and one sample of sharp-tooth snapper exceeded the guidelines by US EPA (1997) and by the Joint FAO/WHO Expert Committee on Food Additives (JECFA 2003), respectively, which indicates that some fish specimens examined (9% and 3% for US EPA and JECFA guidelines, respectively) were hazardous for consumption at the ingestion rate of Cambodian people (32.6 g/d). It is suggested that fish is probably the main source of Hg for Cambodian people. However, extremely high Hg concentrations were observed in some individuals and could not be explained by Hg intake from fish consumption, indicating other contamination sources of Hg in Cambodia.

The level of Hg in human hair of Tanzania residents, especially around Lake Victoria (103 fisherman and families) was performed by Harada et al.

(1999). One fisherman and three women living in the Seweya fishing village showed high THg level of 48 μg/g, near to 50 μg/g, and greater in the head hair (highest value, 416 μg/g). THg levels in the head hair from other fishermen and their families living on the coast of the Lake Victoria were 1.00–3.79 μg/g on average (two cases with values >10 μg/g: 17.4 and 37.9 μg/g). Also, the THg level in head hair from fishermen in Hombolo was 1.00 μg/g on average. In addition, THg contents in fish caught in Lake Victoria were 63 ppb in tilapia, 12 ppb in freshwater sardine, and 8.9 ppb in catfish. The three female subjects with high Hg levels had made habitual use of toilet soap containing considerable amounts of Hg for several years, whereas the fisherman with a high Hg level of 70.4 μg/scarcely used it.

H. Organic Mercury Poisoning in Iraq

Epidemics of organic Hg poisoning from consumption of grain treated with organomercurial fungicides have also occurred in Iraq and Guatemala. By the time the severe Iraq outbreak occurred in 1971, epidemiologists and toxicologists were alerted and analytical results (mainly hair Hg) were obtained and used in risk assessment (Amin-Zaki et al. 1974); this resulted in calculation by the US Environmental Protection Agency (US EPA 1997) of a reference dose of 0.3 mg/kg/d for adults. More recently, supported by a comprehensive review by the National Research Council (NRC 2000), EPA has developed a reference dose of 0.1 mg/kg/d, sufficient to protect the most sensitive endpoint, i.e., fetal neurobehavioral development. Hair Hg was the main biomarker.

VII. Conclusions

These cases cited here illustrate exposure to both elemental and organic Hg compounds in air and food, uptake through lungs and skin, and elevations of Hg in blood, urine (Weiner and Nylander 1995), liver, and hair (Evans et al. 1998). The most useful biomarker depends on the species of metal (e.g., hair is useful for organic but less useful for elemental exposure), and whether one seeks, current or historic exposure information. Identification of the metal species, its bioavailability in the medium, and the levels of a biomarker can allow estimates of or validate risk assessments.

Summary

Hair has been used in many studies as a bioindicator of mercury exposure for human populations. At the time of hair formation, mercury from the blood capillaries penetrates into the hair follicles. As hair grows approximately 1 cm each month, mercury exposure over time is recapitulated in hair strands. Mercury levels in hair closest to the scalp reflect the most recent exposure, while those farthest from the scalp are representative of previous blood concentrations. Sequential analyses of hair mercury have been useful for identifying seasonal variations over time in hair mercury content, which may be the result of seasonal differences in bioavailability of fish and differential consumption of piscivorous and herbivorous fish species. Knowledge of the relation between fish-eating practices and hair mercury levels is particularly important for adequate mitigation strategies.

Methyl mercury is well absorbed, and because the biological half-life is long, the body burden in humans may reach high levels. People who frequently eat contaminated seafood can acquire mercury concentrations that are potentially dangerous to the fetus in pregnant women. The dose–response relationships have been extensively studied, and the safe levels of exposure have tended to decline. Individual methyl mercury exposure is usually determined by analysis of mercury in blood and hair. The objective of the present review was to examine variations in hair mercury levels from different populations with respect to fish-eating practices.

References

Abe T, Ohtsuka R, Hongo T, Suzuki T, Tohyama C, Nakano A, Akagi H, Akimichi T (1995) High hair and urinary mercury levels of fish eaters in the nonpolluted environment of Papua New Guinea. Arch Environ Health 50:367–373.

Agusa T, Kunito T, Iwata H, Monirith IN, Tanca TS, Subramanian A, Tanabe S (2005) Mercury contamination in human hair and fish from Cambodia: levels, specific accumulation and risk assessment. Environ Pollut 134:79–86.

Airey D (1983a) Mercury in human hair due to environment and diet: a review. Environ Health Perspect 52:303–316.

Airey D (1983b) Total mercury concentrations in human hair from 13 countries in relation to fish consumption and location. Sci Total Environ 31:157–180.

Akagi H, Malm O, Kinjo Y, Harada M, Branches FJP, Pfeiffer WC, Kato H (1995) Methylmercury pollution in Amazon, Brazil. Sci Total Environ 175:85–96.

Akagi H, Grandjean PH, Takizawa Y, Weihe P (1998) Methylmercury dose estimation from umbilical cord concentrations in patients with Minamata disease. Environ Res A 77:98–103.

Al-Majed NB, Preston MR (2000) Factors influencing the total mercury and methylmercury in the hair of the fisherman of Kuwait. Environ Pollut 100:239–250.

Amin-Zaki L, Elhassani S, Majeed MA, Clarkson TW, Doherty RA, Greenwood M (1974) Intra-uterine methylmercury poisoning in Iraq. Pediatrics 54(5):587–595.

Amorim MI, Mergler D, Bahia MO, Dubeau H, Miranda D, Lebel J, Burbano RR, Lucotte M (2000) Cytogenetic damage related to low levels of methyl mercury contamination in the Brazilian Amazon. An Acad Bras Cienc 72(4):497–507.

Ask Bjornberg K, Vahter M, Petersson-Grawe K, Glynn A, Cnattingius S, Darnerud PO, Atuma S, Aune M, Becker M, Berglund M (2003) Methylmercury and inorganic mercury in Swedish pregnant women and in cord blood: influence of fish consumption and dental amalgam. Environ Health Perspect 111:637–641.

Ask Bjornberg K, Vahter M, Petersson-Grawe K, Berglund M (2005) Methylmercury exposure in Swedish women with high fish consumption, Sci Total Environ 341:45–52.
ATSDR (1997) Top 20 Hazardous Substances: ATSDR/EPA Priority List for 1995. http://atsdrl.atsdr.cdc.gov:8080/cxcx3.html.
ATSDR (2000) Toxicologic Profile for Mercury. Agency for Toxic Substances and Disease Registry, Centers for Disease Control and Prevention, Atlanta, GA.
Barbosa AC, Dórea JG (1998) Indices of mercury contamination during breast feeding in the Amazon Basin. Environ Toxicol Pharmacol 6:71–79.
Barbosa AC, Boischio AA, East GA, Ferrari I, Gonçalvez A, Silvia PRM, Da Cruz TME (1995) Mercury contamination in the Brazilian Amazon. Environmental and occupational aspects. Water Air Soil Pollut 80:109–121.
Barbosa AC, Silva SRL, Dorea JG (1998) Concentration of mercury in hair of indigenous mothers and infants from the Amazon Basin. Arch Environ Contam Toxicol 34:100–105.
Batista J, Schuhmacher M, Domingo J, Corbella J (1996) Mercury in hair for a child population from Tarragona Province, Spain. Sci Total Environ 193:143–148.
Bencko V (1995) Use of human hair as a biomarker in the assessment of exposure to pollutants in occupational and environmental settings. Toxicology 101:29–39.
Bidone ED, Castilhos ZC, Cid de Souza TM, Lacerda LD (1997) Fish contamination and human exposure to mercury in the Tapajos riverine basin, Para state, Amazon, Brazil: a screening approach. Bull Environ Contam Toxicol 59:194–201.
Blayney MB (2001) The need for empirically derived permeation data for personal protective equipment: the death of Dr. Karen E. Wetterhahn. Appl Occup Environ Hyg 16 (2):233–236.
Bloom NS (1992) On the chemical form of mercury in edible fish and marine invertebrate tissue. Can J Fish Aquat Sci 49:1010–1017.
Boischio AAP, Cernichiari E (1998) Longitudinal hair mercury concentration in riverside mothers along the upper Madeira. River Brazil. Environ Res 77:79–83.
Boischio AAP, Henshel DS (1996) Risk assessment of mercury exposure through fish consumption by the riverside people in the Madeira basin, Amazon, 1991. Neurotoxicology 17(1);169–176.
Bou-Olayan A, Al-Yakoob S (1994) Mercury in human hair: a study of residents in Kuwait. J Environ Sci Health A29: 1541–1551.
Budtz-Jørgensen E, Keiding N, Grandjean P, Weihe P, White RF (2003) Consequences of exposure measurement error for confounder identification in environmental epidemiology. Stat Med 22:3089–3100.
Buzina R, Stegnar P, Buzina-Suboticanec K, Horvat M, Petric I, Farley TMM (1995) Dietary mercury intake and human exposure in an Adriatic population. Sci Total Environ 170:199–208.
Carroll RJ (1998) Measurement error in epidemiologic studies. In: Armitage P, Colton T (eds) Encyclopedia of Biostatistics. Wiley, Chichester, pp 2491–2519.
Cernichiari E, Toribara T, Liang L, Marsh D, Berline M, Myers G, Cox C, Shamlaye C, Choisy O, Davidson P, Clarkson T (1995) The biological monitoring of mercury in the Seychelles study. Neurotoxicology 16:613–628.
Chang LW (1996) Toxicology of Metals. Boca Raton, Lewis, FL.
Choi BH, Lapham LW, Amin-Zaki L, Saleem T (1978) Abnormal neuronal migration, deranged cerebral cortical organization, and diffuse white matter

astrocytosis of human fetal brain: a major effect of methylmercury poisoning in utero. J Neuropathol Exp Neurol 37(6):719–733.
Clarkson TW (1997) The toxicology of mercury. Crit Rev Clin Lab Sci 34:369–403.
Clarkson TW, Friberg L, Nordberg GF, Sager PR (1988) Biological monitoring of toxic metals. Plenum Press, New York.
Craig PJ (ed) (1986) Organometallic Compounds in the Environment: Principles and Reactions. Longman, Leicester, UK, pp 65–101.
Dickman MD, Leung CKM, Leong MKH (1998) Hong Kong male subfertility links to mercury in human hairs and fish. Sci Total Environ 214:165–174.
Dixon R, Jones B (1994) Mercury concentrations in stomach contents and muscle of five species from the North East Coast of England. Mar Pollut Bull 28:741–745.
Dolbec J, Mergler D, Sousa Passos CJ, Sousa de Morais S, Lebel J (2000) Methylmercury exposure affects motor performance of a river in the population of the Tapajos River, Brazilian Amazon. Int Arch Occup Environ Health 73(3):195–203.
Dolbec J, Mergler D, Larribe F, Roulet M, Lebel J, Lucotte M (2001) Sequential analysis of hair mercury levels in relation to fish diet of an Amazonian population, Brazil. Sci Total Environ 271(1–3):87–97.
Duffy LK, Kaiser C, Ackley C, Richter KS (2001) Mercury in the hair of large Alaskan herbivores: routes of exposure. Alces 37:293–301.
Duffy LK, Duffy RS, Finstad G, Gerlach C (2005) A note on mercury levels in the hair of Alaskan reindeer. Sci Total Environ 339:273–276.
Ebinghaus R, Hintelmann H, Wilken RD (1994) Mercury cycling in surface waters and in the atmosphere: species analysis for the investigation of transformation and transport properties of mercury. Fresenius J Anal Chem 350:21–29.
Egeland GM, Middaugh JP (1997) Balancing fish consumption benefits with mercury exposure. Science 278:1904–1905.
Egeland GM, Feyk LA, Middaugh JP (1998) Use of traditional foods in a healthy diet in Alaska: risks in perspective. State Alaska Epidemiol Bull. Recommend Rep 2(1):140.
Eto K (2000) Minamata disease. Neuropathology (suppl) 20:S14–S19.
Evans RD, Addison EM, Villeneuve JY, MacDonald KS, Joachim DG (1998) An examination of spatial variation in mercury concentrations in otter (*Lutra canadensis*) in south-central Ontario. Sci Total Environ 213:2399–245.
Fiedler N, Udasin I, Gochfeld M, Buckler G, Kelly McNeil K, Kipen H (1999) Neuropsychological and stress evaluation of a residential mercury exposure. Environ Health Perspect 107(5):343–347.
Fitzgerald W (1995) Is mercury increasing in the atmosphere? The need for an atmospheric mercury network (AMNET). Water Air Soil Pollut 80(1):245–254.
Fleming L, Watkins S, Kaderman R, Levin B, Ayyar D, Bizzio M, Stephens D, Bean J (1995) Mercury exposure in humans through food consumption from the Everglades of Florida. Water Air Soil Pollut 80:41–48.
Foo SC, Tan TC (1998) Elements in the hair of Southeast Asian islanders. Sci Total Environ 209:185–192.
Foo SC, Ngim CH, Phoon WO, Lee J (1988) Mercury in scalp hair of healthy Singapore residents. Sci Total Environ 72:113–122.
Foo SC, Khoo NY, Heng A, Chua LH, Chia SE, Ong CN, Ngim CH, Jeyaratnam J (1993) Metals in hair as biological indices for exposure. Int Arch Occup Environ Health 65(1 suppl):S83–S86.

Futatsuka M, Kitano T, Shono M, Fukuda Y, Ushijima K, Inaoka T, Nagano M, Wakamiya J, Miyamoto K (2000) Health surveillance in the population living in a methylmercury polluted area over a long period. Environ Res 83(2):83–92.

Galster WA (1976) Mercury in Alaskan Eskimo mothers and infants. Environ Health Perspect 15:135–140.

Grandjean P (1984) Lead poisoning: hair analysis shows the calendar of events. Hum Toxicol 3:223–228.

Grandjean P, Weihe P, Nielsen JB (1994) Methylmercury: significance of intrauterine and postnatal exposures. Clin Chem 40:1395–1400.

Grandjean P, Budtz-Jørgense, E, White RF, Jørgensen PJ, Weihe P, Debes F, Keiding N (1999) Methylmercury exposure biomarkers as indicators of neurotoxicity in 7-year-old children. Am J Epidemiol 150:301–305.

Grandjean P, Jørgensen PJ, Weihe P (2002) Validity of mercury exposure biomarkers. In: Wilson SH, Suk WA (eds) Biomarkers of Environmentally Associated Disease. CRC Press/Lewis, Boca Raton, FL, pp 235–247.

Guimaraes JR, Roulet M, Lucotte M, Mergler D (2000) Mercury methylation along a lake-forest transect in the Tapajos river floodplain, Brazilian Amazon: seasonal and vertical variations. Sci Total Environ 261(1–3):1–98.

Hać E, Krzyżanowski M, Krechniak J (2000) Total mercury in human renal cortex, liver, cerebellum and hair. Sci Total Environ 248:37–43.

Harada M (1995) Minamata disease: methylmercury poisoning in Japan caused by environmental pollution. Crit Rev Toxicol 25:1–24.

Harada M, Nakaniski J, Konuma S, Ohno K, Kimura T, Yanaguchi H, Tsuruta K, Kizaki T, Ookawara T, Ohno H (1998) The present mercury contents of scalp hair and clinical symptoms in inhabitants of the Minamata area. Environ Res A 77:160–164.

Harada M, Nakachi S, Cheu T, Hamada H, Ono Y, Tsuda T, Yanagida K, Kizaki T, Ohno H (1999) Monitoring of mercury pollution in Tanzania: relation between head hair mercury and health. Sci Total Environ 227:249–256.

Holsbeek L, Das HK, Joris CR (1996) Mercury in human hair and relation to fish consumption in Bangladesh. Sci Total Environ 189:181–188.

Hörsted-Bindslev P (2004) Amalgam toxicity—environmental and occupational hazards. J Dent 32:359–365.

Huel G, Everson RB, Menger I (1984) Increased hair cadmium in newborns of women occupationally exposed to heavy metals. Environ Res 35:115–121.

JECFA (2003) Summary and Conclusions of the 61st Meeting of the Joint FAO/WHO Expert Committee on Food Additives (JECFA). JECFA/61/SC. FAO, Rome, Italy.

Kelman BJ, Walter BK, Sasser LB (1982) Fetal distribution of mercury following introduction of methylmercury into porcine maternal circulation. J Toxicol Environ Health 10:191–200.

Kondo K (2000) Congenital Minamata disease: warnings from Japan's experience. J Child Neurol 2000;15(7):458–464.

Kurland LT, Faro SN, Siedler H (1960) Minamata disease. World Neurol 1(4):370–395.

Kyle J, Ghani N (1983) Mercury concentrations in canned and fresh fish and its accumulation in a population of Port Moresby residents. Sci Total Environ 26:157–162.

Lind B, Friberg L, Nylander M (1988) Preliminary studies on methylmercury biotransformation and clearance in the brain of primates. II. Demethylation of mercury in brain. J Trace Elem Exp Med 1:49–56.
Lindberg A, Ask Björnberg K, Vahter M, Berglund M (2004) Exposure of methylmercury in non-fish-eating people in Sweden. Environ Res 96:28–33.
Lodenius M, Seppaenen A, Herranen M (1983) Accumulation of mercury in fish and man from reservoirs in northern Finland. Water Air Soil Pollut 19:237–246.
López-Artiguez M, Grilo A, Martinez D, Soria ML, Muñez L, Ruano E, Moreno E, García-Fuente F, Repetto M (1994) Mercury and methylmercury in population risk groups on the Atlantic Coast of Southern Spain. Arch Environ Contam Toxicol 27:415–419.
Malm O, Pfeiffer WC, Souza CMM, Reuther R (1990) Mercury pollution due to gold mining in the Madeira river basin, Brazil. Ambio 19:11–15.
Malm O, Branches F, Akagi H, Castro M, Pfeiffer W, Harada M, Bastos W, Kato H (1995) Mercury and methylmercury in fish and human hair from the Tapajos river basin, Brazil. Sci Total Environ 175:141–150.
Marsh D, Turner M, Smith J, Allen P, Richale V (1995) Fetal methylmercury study in Peruvian fish-eating population. Neurotoxicology 16:717–726.
Mason RP, Reinfelder JR, Morel FMM (1995) Bioaccumulation of mercury and methylmercury. Water Air Soil Pollut 80:915–921.
Matsubara J, Machida K (1985) Significance of elemental analysis of hair as a means of detecting environmental pollution. Environ Res 38:225–238.
Matsumoto H, Koya G, Takeuchi T (1965) Fetal Minamata disease: A neuropathological study of two cases of intrauterine intoxication by a methyl mercury compound. J Neuropathol Exp Neurol 24:563–574.
Matsuo N, Suzuki T, Akagi H (1989) Mercury concentration in organs of contemporary Japanese. Arch Environ Health 44:298–303.
Mempel M, Chau YK, Dutka BJ, McInnis R, Kwan KK, Liu D (1995) Toxicity of organomercury compounds: bioassay results as a basis for risk assessment. Analyst 120:721–724.
Moreiras O, Cuadrado C (1992) Theoretical study of the intake of trace elements (nutrients and contaminations) via total diet in some geographical areas of Spain. Biol Trace Elem Res 32:93–103.
Mortada WI, Sobh MA, El-Defrawy MM, Farahat SE (2002) Referene intervals of cadmium, lead and mercury in blood, urine, hair, and nails among residents in Mansoura City, Nile delta, Egypt. Environ Res A 90:104–110.
Moszczynski P, Moszczynski P (1990) Current views on biotransformation and metabolism of mercury (in Polish). Post Hig Med Dośw 44:153–180.
Mueller KA, Matz AC (2002) Water quality, and metal and metalloid concentrations in water, sediment, and fish tissues from Innoko National Wildlife Refuge, Alaska, 1995–1997. Tech Rep NAES-TR-02-01. U.S. Fish and Wildlife Service, Northern Alaska Ecological Services, Fairbanks.
Mueller KA, Snyder-Conn E, Doyle T (1993) Contaminant baseline data for water, sediments, and fish of Selawik National Wildlife Refuge, Alaska, 1987–1988. Tech Rep NAES-TR-93-02. U.S. Fish and Wildlife Service, Northern Alaska Ecological Services, Fairbanks.
Mueller KA, Snyder-Conn E, Bertram M (1996) Water quality, and metal and metalloid contaminants in sediments and fish of Koyukuk, Nowitna, and the northern unit of Innoko National Wildlife Refuges, Alaska, 1991. Tech Rep

NAES-TR-96-01. US Fish and Wildlife Service, Northern Alaska Ecological Services, Fairbanks.
Muramatsu Y, Parr RM (1988) Concentrations of some trace elements in hair, liver and kidney from autopsy subjects: relationship between hair and internal organs. Sci Total Environ 76:29–40.
Muszyńska-Zimna E (1982) Total mercury content in the hair and urine of individuals inhabiting the region of Łódź and being not professionally exposed to the said element (in Polish). Bromat Chem Toksykol 10:127–128.
Nagakawa R (1995) Concentration of mercury in hair of Japanese people. Chemosphere 30:127–133.
New Jersey Mercury Task Force (2001) Mercury in New Jersey. New Jersey Department of Environmental Protection, Trenton, NJ.
Nierenberg DW, Nordgren RE, Chang MB, Siegler RW, Blayney MB, Hochberg F, Toribara TY, Cernichiari E, Clarkson T (1998) Delayed cerebellar disease and death after accidental exposure to dimethylmercury. N Engl J Med 338(23): 1672–1676.
Nowak B (1996) Occurrence of heavy metals, sodium, calcium and potassium in human hair, teeth, and nails. Biol Trace Elem Res 52:11–22.
Nylander M, Friberg L, Lind B (1987) Mercury concentrations in the human brain and kidneys in relation to exposure from dental amalgam fillings. Swed Dent J 11:179–187.
NRC (2000) Toxicological Effects of Methylmercury. National Research Council/National Academy Press, Washington, DC.
Ohno H, Doi R, Tani Y, Harada M (1984) Mercury content of head hair from residents on the coast of Jakarta Bay. Bull Environ Contam Toxicol 33:282–285.
Olivero J, Mendoza C, Mestre J (1995) Hair mercury levels in different occupational groups in southern Bolivar (Columbia). Rev Sande–Publ SAO Paulo 29: 376–379.
Ong CN, Chia SE, Foo SC, Ong HY, Tsakok M, Liouw P (1993) Concentrations of heavy metals in maternal and umbilical cord blood. BioMetals 6:61–66.
Orloff KG, Ulirsch G, Wilder L, Block A, Fagliano J, Pasqualo J (1997) Human exposure to elemental mercury in a contaminated residential building. Arch Environ Health 52(3):169–172.
Phelps RW, Clarkson TW, Kershaw TG, Wheatly B (1980) Interrelationships of blood and hair concentrations in North American population exposed to methylmercury. Arch Environ Health 35:161–167.
Philip O, Oyuah MD (2000) Mercury poisoning. Curr Probl Pediatr 30:91–99.
Ponce RA, Bartell SM, Kavanagh TJ, Woods JS, Griffith WC, Lee RC, Takaro TK, Faustman EM (1998) Uncertainty analysis methods for comparing predictive models and biomarkers: a case study of dietary methylmercury exposure. Reg Toxicol Pharmacol 28:96–105.
ROPME (1988) Regional Organization for the Protection of the Marine Environment. Results of project number: KA/5102 82 10(2363) rev.4. Survey of mercury in fish and sediments from the ROPME Sea Area (unpublished report). Prepared by IAEA-MESL, Monaco.
Roulet M, Lucotte M, Farella N, Serique G, Coelho H, Passos CJS, de Jesus da Silva E, de Andrade PS, Mergler D, Guimaraes JR, Amorim M (1999) Effects of recent human colonization on the presence of mercury in Amazonian ecosystems. Water Air Soil Pollut 112(3–4):297–313.

Ryozo N (1995) Concentrations of mercury in hair of diseased people in Japan. Chemosphere 30:135–140.

Santos ECO, Jesus IM, Brabo ES, Loureiro ECB, Mascarenhas AF, Weirich J, Câmara VM, Cleary D (2000) Human health in riverside Amazon communities living in exposed and not exposed areas to mercury pollution from gold mining. Environ Res 86:100–107.

Santos ECO, Jesus IM, Câmara VM, Brabo ES, Loureiro ECB, Mascarenhas AF, Weirich J, Luiz RR, Cleary D (2002) Mercury exposure in Munduruku from the community of Sai Cinza, State of Pará, Brazil. Environ Res 90:98–103.

Sarmani S, Koshy T, Zakaria Z (1985) Trace element analysis of human hair samples from rural and urban population of Selangor, Malaysia. In: Koh LL, Hew CS (eds) Proceedings of the Third Symposium on our Environment, 27–29 March 1984, Singapore. Faculty of Science, National University of Singapore, pp 473–477.

Sarmani SB, Kiprawi AZ, Ismail RB (1994) Mercury determination in hair of Malaysian fishermen by neutron activation analysis. Biol Trace Elem Res 43–45:435–441.

Schuhmacher M, Batista J, Bosque MA, Domingo JL, Corbella J (1994) Mercury concentrations in marine species from the coastal area of Tarragona Province, Spain. Dietary intake of mercury through fish and seafood consumption. Sci Total Environ 156:269–273.

Segade SR, Tyson J (2003) Determination of inorganic mercury and total mercury in biological and environmental samples by flow injection-cold vapor-atomic absorption spectrometry using sodium borohydride as the sole reducing agent. Spectrochim Acta Part B 58:797–807.

Seppänen K, Kantola M, Lattikainen R, Nyyssönen K, Valkonen V-P, Salomen JT (2000) Effect of supplementation with organic selenium on mercury status as measured by in public hair. J Trace Elem Med Biol 14:84–87.

Smith JC, Farris FF (1996) Methyl mercury pharmacokinetics in man: a reevaluation. Toxicol Appl Pharmacol 137:245–252.

SNFA (Swedish National Food Administration) (2002) Kostråd och näringsrekommendationer (in Swedish). Swedish National Food Administration, Uppsala, Sweden. http://www.sl.se (Accessed March 2002.)

Srikumar TS, Ockerman PA, Akesson B (1992) Trace element status in vegetarians from southern India. Nutr Res 12:187–198.

Srikumar TS, Kaligard A, Lindeberg S, Ockerman PA, Akesson B (1994) Trace element concentrations in hair of subjects from two South Pacific Islands Atafu Tokelau and Kitava Papua New Guinea. J Trace Elem Electrolytes Health Dis 8:21–26.

Srogi K (2004) Concentrations of cadmium and copper in the natural medicinal plants. Bromatol Chem Toksykol 37(3):291–294.

Srogi K (2005) Determination of cadmium and lead in the hairs of people from the Silesia region. Bromatol Chem Toksykol 38(1):67–74.

Stern AH, Smith AE (2003) An assessment of the cord blood: maternal blood methylmercury ratio: implications for risk assessment. Environ Health Perspect 111:1465–1470.

Stern AH, Gochfeld M, Weisel C, Burger J (2001) Mercury and methylmercury exposure in the New Jersey pregnant population. Arch Environ Health 56(1):4–10.

Stewart CP, Stolman A (1961) Toxicology. Mechanisms and Analytical Methods, vol II. Academic Press, New York.

Suzuki T (1988) Hair and nails: advantages and pitfalls when used in biological monitoring. In: Clarkson TW, Friberg L, Nordgerg GF, Sager PR (eds) Biological Monitoring of Toxic Metals. Plenum Press, New York, pp 623–641.

Suzuki T, Yamamoto R (1982) Organic mercury levels in human hair with and without storage for eleven years. Bull Environ Contam Toxicol 28:186–188.

Suzuki T, Hongo T, Yoshinaga J, Imaitt, Nakazawa M, Matsuo N, Akagi H (1993) The hair–organ relationship in mercury concentration in contemporary Japanese. Arch Environ Health 48:221–229.

Swartout J, Rice G (2000) Uncertainty analysis of the estimated ingestion rates used to derive the methylmercury reference dose. Drug Chem Toxicol 23:293–306.

Takeuchi T, Hayashi T, Takada J (1982) Variation of elemental concentration in hair of the Japanese in term of age, sex and hair treatment. J Radioanal Chem 70:29–55.

Tsuchiya H, Mitani K, Kodama K, Nakata T (1984) Placental transfer of heavy metals in normal pregnant Japanese women. Arch Environ Health 39:11–17.

Turner MD, Marsh DO, Smith JC, Inglis JB, ClarksoN TW, Rubio CE, Chiriboga J, Chiriboga CC (1980) Methylmercury in populations eating large quantities of marine fish. Arch Environ Health 35:367–378.

US EPA (Environmental Protection Agency) (1997) Report to Congress: Mercury. U.S. Environmental Protection Agency, Washington, DC.

US EPA (2001) Water Quality Criterion for the Protection of Human Health: Methylmercury. Final. EPA-823-R-01-001. Washington, DC. http://www.epa.gov/waterscience/criteria/methylmercury/document.html (accessed August 15, 2003).

Valentino L, Torregrossa M, Saliba L (1995) Health effects of mercury ingested through consumption of sea food. Pollut Medit Sea 32:41–47.

Watt J (2001) Mercury poisoning victims could increase by 20,000. Lancet 358:1349.

Weihe P, Grandjean P, Debes F, White R (1996) Health implications for Faroe Islanders of heavy metals and PCBs from pilot whales. Sci Total Environ 186:141–148.

Weiner JA, Nylander M (1995) An estimation of the uptake of mercury from amalgam fillings based on urinary excretion of mercury in Swedish subjects. Sci Total Environ 168:255–265.

Wheeler M (1996) Measuring mercury. Environ Health Perspect 104:826–830.

Wiadrowska B, Syrowatka T (1983) Assessment of total mercury content in human tissues. II. Mercury content of human hairs in the general population and in subjects with occupational exposure to mercury vapours. Roczn PZH 34:87–94.

Wiadrowska B, Syrowatka T, Mazurek J (1983) Evaluation of total mercury content of human tissues III. Mercury content of human kidneys and brain in the general population. Roczn PZH 34:171–174.

World Health Organization (WHO) (1989) Toxicological evaluation of certain food additives and contaminants. In: WHO Food Additives Series. Cambridge University Press, Cambridge, UK, pp 295–328.

World Health Organization (WHO) (1990) International Programme on Chemical Safety. Environmental Health Criteria 101: Methylmercury. World Health Organization, Geneva.

Yamaguchi S, Matsumoto H, Kaku S, Tateishi M, Shiramizu M (1975) Factors affecting the amount of mercury in human scalp hair. Am J Public Health 1975; 65:484–488.

Yamamoto R, Suzuki T (1978) Effects of artificial hair-waving on hair mercury values. Int Arch Occup Environ Health 42:1–9.

Yasutake A, Matsumoto M, Yamaguchi M, Hachiya N (2003) Current hair mercury levels in Japanese: survey in five districts. Tohoku J Exp Med 199:161–169.
Zaki R, Abdulraheem M (1973) Distribution of trace metals in the marine environment of Kuwait (final data report). Ministry of Public Health, Kuwait.

Manuscript received April 6; accepted April 13, 2006

Toxicology of 1,3-Butadiene, Chloroprene, and Isoprene

Harrell E. Hurst

Contents

I. Introduction 131
II. Butadiene 132
A. Chemical Information and Use 132
B. Toxicity 134
C. Metabolism and Kinetics 137
D. Biomarkers 142
E. Genotoxicity 145
F. Human Epidemiology 148
G. Risk Assessment 150
III. Chloroprene 150
A. Chemical Information and Use 150
B. Toxicity 151
C. Metabolism and Kinetics 152
D. Biomarkers 155
E. Genotoxicity 155
F. Human Epidemiology 156
G. Risk Assessment 157
IV. Isoprene 158
A. Chemical Information and Use 158
B. Toxicity 159
C. Metabolism and Kinetics 160
D. Biomarkers 162
E. Genotoxicity 162
F. Human Epidemiology 164
Summary 164
References 165

I. Introduction

The diene monomers 1,3-butadiene, chloroprene, and isoprene, shown in Fig. 1, are important building blocks used in the synthesis of polymers. These volatile organic compounds contribute significantly to the atmospheric burden of organic chemicals, appearing at low parts per billion (ppb)

Communicated by David M. Whitacre.

H.E. Hurst (✉)
Department of Pharmacology and Toxicology, University of Louisville School of Medicine, Louisville, KY 40292, U.S.A.

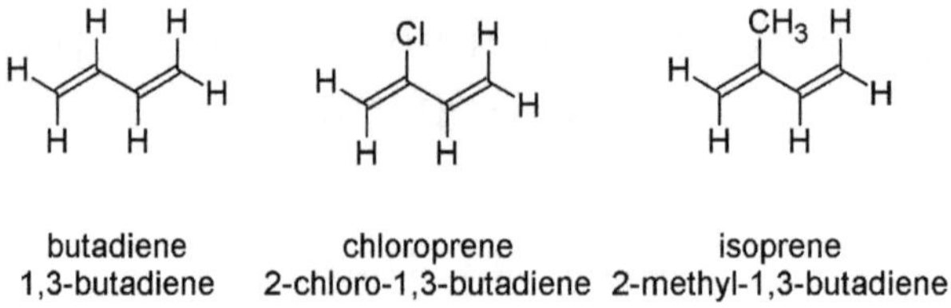

Fig. 1. Structures of butadiene, chloroprene, and isoprene.

concentrations in ambient air near industrial sites where they are used. Occupational exposures to these dienes occur at higher air concentrations approaching or slightly exceeding the parts per million (ppm) threshold in facilities that produce or polymerize these monomers. Such exposures present significant continuing concern to occupational hygienists and epidemiologists in the quest to provide safe working conditions during production. These three chemicals, which differ chemically only by substitution of a hydrogen, chlorine, or methyl group at the 2-carbon of the molecule, provide interesting comparisons among their physical properties, occurrence, uses, and potential health effects. This review does not attempt comprehensive review of these chemicals, as that would fill many volumes. Search of the National Library of Medicine Medline database yielded more than 1,100, 700, and 130 citations, respectively, when butadiene, isoprene, and chloroprene were sought as keywords indexed from publications between 1966 and October 2005. Rather, this chapter attempts to summarize current knowledge of toxic effects from existing primary research reports and reviews, to illustrate their metabolism and proposed mechanisms of toxicity, and to draw parallels and illustrate significant differences among these dienes. The hope is that such a summary will provide perspective on potential hazardous or safe uses of these important materials. Illustration of similarities and differences among these small molecules should intrigue thought and may provoke work toward additional understanding of the complexities that comprise interactions among these deceptively simple chemicals, the environment, and living systems.

II. Butadiene

A. Chemical Information and Use

1,3-Butadiene (CAS no. 106-99-0), known also as α,γ-butadiene, biethylene, bivinyl, butadiene, butadiene-1,3, divinyl, erythrene, and vinylethylene, is a colorless gas that has a mild gasoline-like odor. Other physical properties are given in Table 1. Butadiene is produced during ethylene production by steam cracking of petroleum, during which petroleum distillates are combined with steam and subjected to high temperature. In this process, a series of condensation and hydrogen transfer reactions occur, and the resulting

Table 1. Physical properties of the dienes, butadiene, chloroprene, and isoprene.

Property	Butadiene	Chloroprene	Isoprene
Standard condition state	Gas	Liquid	Liquid
Molecular weight	54.09	88.54	68.13
Boiling point (°C)	–4.41	59.4	34.1
Melting point (°C)	–108.9	–130.0	–146.0
Specific gravity	0.65[a]	0.96	0.68
Convert[b] 1 ppm to mg/m^3	2.21	3.62	2.79
Convert[c] 1 mg/m^3 to ppm	0.452	0.276	0.359

[a]At –4.4°C; others at 20°C.
[b]Multiply ppm by given factor.
[c]Multiply mg/m^3 by given factor.
Source: Merck Index (1996); Yaws (1976).

decomposition and rearrangement products are separated and collected. Butadiene, a major commodity of petroleum refining, is 1 of the 20 industrial chemicals produced in greatest volume. More than 5,000,000 t were produced worldwide in 2004 (CEN 2005). Of that total, approximately equal amounts (43%) were produced in Europe and the United States, whereas proportions of 8% and 6% were produced in Japan and Canada, respectively. Butadiene is used in production of a wide variety of rubber, latex, and plastic compounds and serves as a key building block for syntheses of other polymers. Among the important derived polymers are styrene-butadiene rubber, which is used in automobile and truck tires; adiponitrile; a key ingredient in the manufacture of nylon; styrene-butadiene latexes; and acrylonitrile-butadiene-styrene plastics, which are durable under high impact. Butadiene polymerizes easily, is highly reactive, and dimerizes to 4-vinylcyclohexene (Lynch 2001a). Butadiene is a significant component [12%–42% (ACC Olefins Panel 2005)] of the Pyrolysis C3+ and C4+ categories listed with the US High Production Volume Chemical Program, a joint effort of the USEPA and the chemical industry to voluntarily fill basic data gaps on high production volume chemicals.

A recent estimate indicated that more than 7,000 workers are exposed to butadiene in the workplace (Lynch 2001a). Data on occupational exposure to butadiene before 1995 have been summarized (Lynch 2001b), indicating previously measured exposures were in the range of 1–3.5 ppm on average, with stated maximum exposures of 10 ppm. Currently the permissible exposure limits (PEL), legally binding limits for worker exposure set by the U.S. Occupational Safety and Health Administration (OSHA), are 1 ppm as an 8-hr time-weighted average and 5 ppm as a short-term exposure limit (NTP 2005a).

Many more persons are exposed to sub- or low ppb levels environmentally. Data published via environmental or state websites indicate ambient air butadiene concentrations just below 1ppb on average (California Air Resources Board 2006; Maryland Department of the Environment 2006; Washington Department of Ecology 2006; WJCCTF 2006), with occasional levels as high as 29ppb near industrial facilities (WJCCTF 2006). Butadiene is a vapor-phase component of tobacco smoke, and approximately 0.4mg per cigarette has been stated to be emitted in environmental tobacco smoke (Penn and Snyder 1996). Butadiene is produced in great quantities by combustion processes that occur in mobile sources (Sapkota et al. 2006), primarily vehicles, and from open burning, such as forest fires and prescribed burns. Emissions from the latter combustion sources have been estimated to comprise more than 90% (61,707t) of the total environmental emissions of butadiene in 1990 (Morrow 2001). For comparison, in the same year estimated releases of butadiene to the environment from various manufacturing processes were about 5.8% (2,350t) of the estimated total environmental release (Morrow 2001). More recent US EPA Toxic Release Inventory data for 2003 (USEPA 2005c) estimate total environmental releases of butadiene by reporting industries at 934t; 60% represented point-source releases into air, 36% were fugitive air emissions, and 3.7% was disposed by underground injection. If these data are comparable to the earlier data of Morrow, industrial emissions have declined during the interval between 1990 and 2003.

Environmental distribution and fate of butadiene largely are determined by atmospheric and photochemical reactions. Distribution of butadiene has been estimated from model calculations that account for physical and chemical properties. Such studies indicate that butadiene partitions between air (99.97%) and water (0.03%), with soil, sediment, suspended sediments, and biota compartments accounting for <0.01% of environmental distribution (ACC Olefins Panel 2005). Photodegradation rates have been estimated from hydroxyl radical photodegradation calculations from which the half-life of butadiene is 1.9hr (ACC Olefins Panel 2005), assuming 12-hr sunlight cycles; this value compares very favorably to the value for benzene (65.8hr) from the same source. Biodegradation by microorganisms has been proposed to proceed via NADH-mediated oxidation to an epoxide intermediate that is hydrolyzed and further oxidized to acetic acid with release of CO_2 (Watkinson and Somerville 1976). The half-life of aerobic aquatic biodegradation has been estimated as ranging from 1 to 4wk (Howard et al. 1991).

B. Toxicity

Butadiene is of significant toxicological interest because of the extent of human exposure in industrial use, its pervasive presence in the atmosphere, and carcinogenic effects noted in chronic rodent inhalation studies. Butadiene is a homologue of vinyl chloride and other substituted ethenes, believed to exhibit carcinogenicity by means of formation of electrophilic epoxide metabolites (Bolt 1986; Gwinner et al. 1983). As a consequence, the toxicological literature on butadiene has been extensively reviewed (Acquavella and Leonard 2001; Bond and Medinsky 2001; Himmelstein et al. 1997; Hughes et al. 2001).

Early studies cited in a very comprehensive review (Himmelstein et al. 1997) noted that high concentrations were anesthetic in animals, as a concentration of 25% was lethal to rabbits within 30 min. Fifty percent of rats and mice succumbed to concentrations of 12.2% and 12.9% after 4- and 2-hr exposures, respectively (Shugaev 1969). These acute exposures produced eye, respiratory tract, and skin irritation, as well as the central nervous system (CNS) effects. Irritation of eyes, nasal passages, throat, and lungs was noted in workers exposed to butadiene during early manufacture of rubber (Wilson 1944).

Studies of toxic effects in animals other than lethality and cancer have been reviewed in detail (Himmelstein et al. 1997). Significant toxic effects noted were biochemical alterations, including depletion of glutathione in liver, lung, and heart with higher inhaled concentrations of butadiene. Depletion of glutathione was noted to be more complete and occurred with lower inhaled concentrations in mice than in rats. This depletion was correlated with increased concentrations of the epoxide metabolites butadiene monoepoxide and butadiene diepoxide (Himmelstein et al. 1995). The latter, more reactive metabolite was measured in mice but not in rats, consistent with greater biochemical toxicity observed in mice.

Morrissey et al. (1990) have presented reproductive and developmental studies sponsored by the U.S. National Toxicology Program (NTP). These studies involved Sprague–Dawley rats and B6C3F1 mice, with mice being more sensitive to butadiene inhalation. Rats exposed to 1,000 ppm gained less weight during pregnancy, but no fetal developmental toxicity was evident. However, Swiss CD-1 mice exhibited developmental toxicity at 200 ppm and 1,000 ppm levels, with anomalies including extra ribs and decreased ossification of sternebrae in the fetuses. Additional NTP studies noted abnormal sperm head morphology in male mice exposed to high (1,000–5,000 ppm) butadiene concentrations, and an increased ratio of dead to total implanted fetuses in dominant lethal assays in female mice. These studies suggested that butadiene, or its metabolites produced at high concentrations in mice, is mutagenic to germ cells. Butadiene has been noted to cause atrophy of gonads in mice (Melnick et al. 1990a). Male B6C3F1 mice exhibited testicular atrophy when exposed to inhaled butadiene at 625 ppm, and females had atrophy of ovaries at concentrations down to 6.25 ppm.

Toxicity of the hematopoietic system is another effect of butadiene in mice exposed to substantial concentrations of butadiene. Macrocytic-megablastic anemia was observed in two strains of male mice in chronic studies (Irons et al. 1986a,b) involving 1,250 ppm butadiene inhalation for 6–24 wk. Alteration of bone marrow stem cell composition and development may also be involved (Leiderman et al. 1986).

Of primary concern is the potential carcinogenicity of butadiene, as indicated by neoplasms noted in multiple organs of rats and mice exposed by inhalation. Chronic studies for carcinogenicity in rodents have been thoroughly reviewed by Himmelstein et al. (1997). In a primary rat study (Owen et al. 1987), butadiene was administered to Sprague–Dawley rats by chronic inhalation at 0, 1,000, or 8,000 ppm by the usual regimen of 6 hr/d, 5 d/wk for 105 wk. Four tumor sites were noted in female rats, but two tumor types, uterine sarcomas and Zymbal gland carcinomas, were not considered as treatment related because incidence was marginal compared to controls. Tumors related to butadiene exposure included mammary adenomas and carcinomas and thyroid follicular cell adenomas; these increased in relation to dose with incidence rates of 50% (control), 79% (1,000 ppm), and 81% (8,000 ppm). In males of the same strain exposed to the same concentrations for 111 wk, pancreatic exocrine adenomas were observed but not considered treatment related, whereas dose-related testicular Leydig cell tumors were noted with incidence rates of 0%, 3%, and 8% for the inhalation groups, respectively. The authors suggested that butadiene was a weak oncogen in rats under conditions of their study.

The NTP has sponsored two chronic inhalation studies to assess butadiene carcinogenicity in mice. The first study exposed B6C3F1 mice to butadiene concentrations of 625 or 1,250 ppm for 60 wk (Huff et al. 1985) but was shortened from the planned 103 wk due to excessive cancer-related mortality. This study indicated acinar cell mammary carcinomas, ovarian granulose cell neoplasms, and hepatocellular neoplasms in females. Cardiac hemangiosarcomas, malignant lymphomas, neoplasms in bronchi and alveoli, and forestomach squamous cell neoplasms were noted in both sexes. Early deaths were caused by lymphocytic lymphomas, and extensive incidence and deaths were believed to mask appearance of tumors in other organs (Melnick et al. 1990b).

A second study (Melnick et al. 1990a) was performed to assess more clearly effects in B6C3F1 mice at lower treatment concentrations, ranging from 6.25 to 625 ppm. Similar multiple tumor sites were seen in both sexes, but females appeared to be more sensitive. Mortality-adjusted tumor rates of controls and treated mice, respectively, are given. At 6.25 ppm females exhibited lung neoplasms in bronchi and alveoli; at 20 ppm, liver neoplasms and lymphomas were noted; at 62.5 ppm, females exhibited adenoma or carcinoma of Harderian gland, neoplasms of ovarian granulosa cell, and mammary carcinoma; and at 200 ppm females exhibited hemangiosarcoma and lymphatic malignant lymphoma. At the highest dose rate, 625 ppm, squamous cell neoplasms of forestomach were seen in the female mice. Males developed corresponding tumors, but at similar adjusted mortality rates at dose levels somewhat higher than did females.

These studies indicated butadiene to be a potent, multisite carcinogen in mice. In comparison with Sprague–Dawley rats, B6C3F1 mice exhibited tumors at 1,000-fold lower doses. The one common tumor site between the species was the mammary gland. In contrast to conclusions reached after the rat study, the authors (Melnick et al. 1990a) sounded an alarm: "So far, no long-term studies on butadiene have been conducted at exposure concentrations that have not shown a carcinogenic response. . . . Because of the correspondence between these animal data and recent epidemiology findings, there is a worldwide public health need to reevaluate current workplace exposure standards for 1,3-butadiene" (Melnick et al. 1990a). These and other studies led OSHA to dramatically reduce the legally binding permissible exposure limit (PEL) for butadiene from 1,000 ppm as an 8-hr time-weighted average (TWA) to 1 ppm effective in February 1997 (OSHA 1996, 2005a). This regulatory action implemented workplace controls on butadiene equivalent to those on vinyl chloride (OSHA 2005b). Under the EPA 1999 *Draft Guidelines for Carcinogen Risk Assessment* (USEPA 1999), 1,3-butadiene was characterized as carcinogenic to humans by inhalation. Although EPA has issued new guidelines for cancer risk assessment (USEPA 2005a), that characterization will remain in effect unless reassessment occurs in the future, according to current EPA policy (USEPA 2005b).

C. Metabolism and Kinetics

A substantial body of literature supports the hypothesis that toxicity of butadiene results from its metabolism to epoxides (Abdel-Rahman et al. 2001; Bolt et al. 1983; Bond et al. 1996; Bond and Medinsky 2001; Boogaard et al. 2004; Elfarra et al. 1996; Himmelstein et al. 1995, 1996; Jackson et al. 2000a,b; Thornton-Manning et al. 1996; Wickliffe et al. 2006). These epoxides are shown in a scheme for butadiene metabolism in Fig. 2, which includes reactions that have been demonstrated in vivo or proposed from in vitro studies. As butadiene is lipophilic, the proportion of the compound that is not exhaled following its inhalation must be converted to products that are more water soluble for excretion by any routes other than the lung. Given the lack of full hydrogen saturation or partially oxidized state of each carbon in butadiene, no "direct hydroxylation" is possible as this requires highly reduced carbon structures. Rather, the initial step is oxidation of butadiene to epoxide metabolites by cytochromes P450 (CYP). Isozymes of this superfamily of monoxygenase enzymes have differing substrate specificities and are largely localized in cellular membranes (Coon 2005). In addition to increasing hydrophilicity to promote excretion, such oxidation is a common mechanism by which unsaturated hydrocarbons are activated to reactive electrophiles (Guengerich 2003; Melnick 2002). The CYP isozyme forms have been characterized, grouped by their amino acid sequence and homology among species, and examined for substrate specificities (Guengerich 1997). Proposed chemical mechanisms for epoxidation of olefins have been reviewed (Guengerich 2003).

Butadiene is oxidized by cytochromes P450 to electrophilic epoxides, including metabolites that contain one or two epoxide (or oxirane) rings

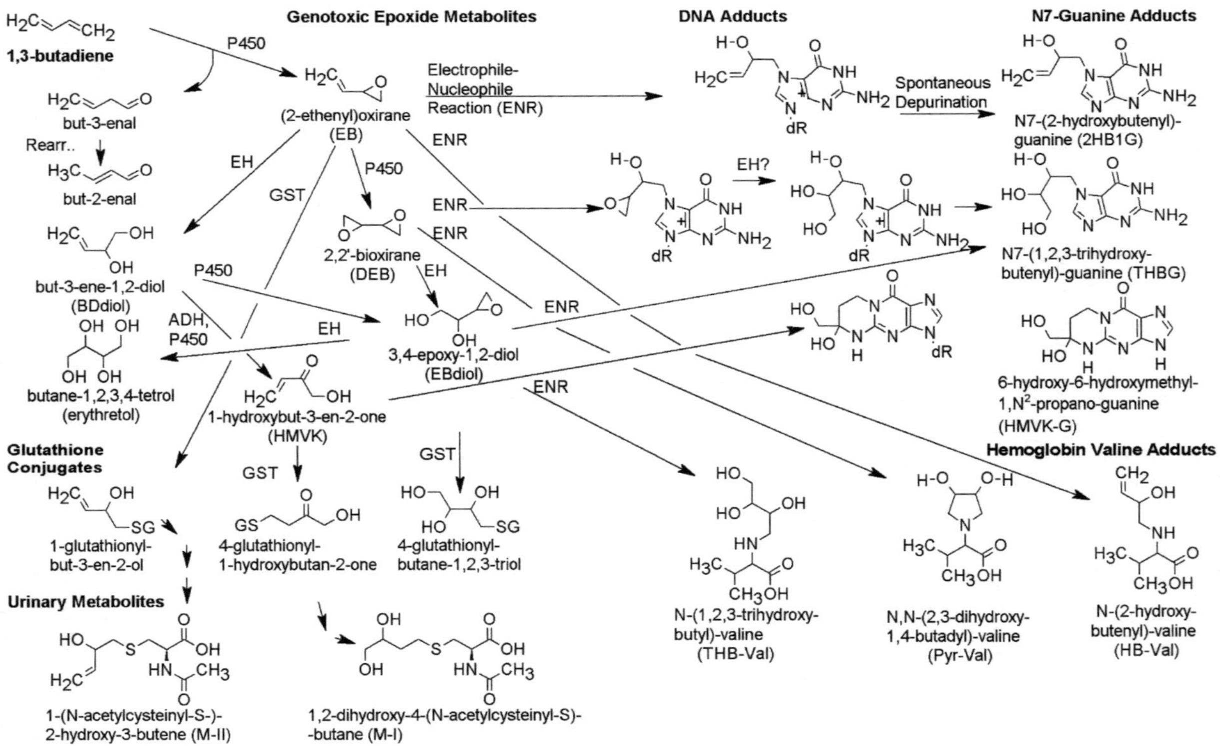

Fig. 2. Metabolic scheme for 1,3-butadiene. *Note*: 1-hydroxy- and 2-hydroxy adduct regiosiomer pairs may exist from electrophile–nucleophile reactions that are not shown for simplicity.

(Bolt et al. 1983; Csanady et al. 1992; Duescher and Elfarra 1994; Himmelstein et al. 1994, 1995; Perez et al. 1997; Thornton-Manning et al. 1995, 1996). Two primary epoxide metabolites are 1,2-epoxy-3-butene (EB), otherwise known as butadiene monoepoxide, epoxybutene, or monoepoxybutene, or 2-ethenyloxirane; and 1,2:3,4-diepoxybutane (DEB), otherwise the diepoxide of butadiene, butane diepoxide or 2,2′-bioxirane (see Fig. 2). The CYP isozymes that metabolize butadiene to its electrophilic epoxides have been determined to include CYP2E1, but probably not exclusively (Jackson et al. 2000b). The human forms that convert butadiene to its monoepoxide appear to be CYP2E1 at low concentration, while CYP2A6 may predominate quantitatively at high substrate concentrations (Elfarra et al. 1996). Additionally, CYP4B1 appears to be a major catalyst of butadiene epoxidation in mouse kidney and liver (Elfarra et al. 2001a). The epoxide metabolites of butadiene are of great significance with respect to the toxicity of butadiene, as these reactive electrophiles have been considered the active mutagenic and carcinogenic species (Elfarra et al. 1996, 2001b; Henderson et al. 2000; Melnick and Huff 1992). Additional epoxide forms, including 3,4,-epoxy-1,2-butanediol (EBdiol), may be involved in the carcinogenic processes (Booth et al. 2004; Melnick 2002; Powley et al. 2005), and is of particular concern as the most abundant genotoxic metabolite in humans. Swenberg and coworkers (Koc et al. 1999) have hypothesized that most of the EBdiol was generated by immediate conversion of DEB to EBdiol within the endoplasmic reticulum. This hypothesis is consistent with the privileged access hypothesis of epoxide hydration (Kohn and Melnick 2000; Melnick 2002).

The relative genotoxic potency of the butadiene epoxide metabolites has been considered to decrease in the order DEB > EB > EBdiol (Jackson et al. 2000a) based on effects seen in rats, mice, and in human cells. Involvement of stereoisomers of DEB, which have not been considered as separate forms in most studies, further complicate understanding (Krause and Elfarra 1997), and the toxicological significance of aldehydes, which are formed in low proportion (2%–5%; Elfarra et al. 1996) relative to EB, are yet to be determined.

The diepoxide, DEB, is considered the most potent genotoxic metabolite due to its capabilities as a bifunctional electrophile and may be the ultimate carcinogenic metabolite of butadiene (Boogaard and Bond 1996; Kligerman et al. 1999). DEB has been found by means of in vitro studies to be formed in greater levels in mice than in rats. Using mouse and rat hepatocytes treated with EB, observed areas under time–concentration curves (AUC) of total DEB that were 8 to 16 fold greater in mice than in rats (Kemper et al. 2001). Increased formation and presence of DEB may explain the greater susceptibility of the mouse over the rat to tumor formation during chronic butadiene exposure (Bond et al. 1996; Bond and Medinsky 2001; Himmelstein et al. 1997), although other studies have suggested a more complex, species-dependent selection of metabolic pathways rather than simple amounts of epoxide metabolites formed for the interspecies difference (Booth et al. 2004; Richardson et al. 1999).

An alternative pathway presents a potentially mutagenic metabolite, hydroxymethylvinyl ketone (HMVK; 1-hydroxybut-3-en-2-one) (Powley et al. 2003). This α,β-unsaturated carbonyl compound has been proposed (Bechtold et al. 1994; Sabourin et al. 1992) as an intermediate in the metabolism of 3-butene-1,2-diol (BDdiol) to 1,2-dihydroxy-4-*N*-acetylcysteinyl)butane, an abundant metabolite (M-I; see Fig. 2) that appears in urine of mammals exposed to butadiene. Although the mechanism of formation has not been fully elucidated, it is likely that HMVK is formed from BDdiol by oxidation via P450 and/or dehydrogenase enzyme(s) with subsequent rearrangement (Kemper et al. 1998; Powley et al. 2003).

Following oxidation, several potential detoxification reactions are possible, as indicated in Fig. 2. Two general types of reaction are important: hydration or hydrolysis of epoxides to vicinal diols via epoxide hydrolase (EH) (Morisseau and Hammock 2005), and formation of glutathione (GSH) conjugates from the electrophilic species mediated by glutathione *S*-transferase (GST) isozymes (Melnick 2002), particularly the GST theta (T1-1) class (Hayes et al. 2005; Landi 2000).

The primary detoxification mechanism for these toxic epoxide metabolites is microsomal epoxide hydrolase (mEH) (Morisseau and Hammock 2005), as the proximity of the membrane-bound hydrolase enzymes to membrane-bound cytochromes P450 may provide "privileged access" and preferential metabolism of epoxides to 1,2-diol metabolites before release into the systemic circulation (Kohn and Melnick 2000; Melnick 2002). A recent publication (Morisseau and Hammock 2005) has reviewed epoxide hydrolases, which are categorized into mEH and soluble EH (sEH). This work stated that no evidence exists that sEH participates in metabolism of xenobiotics in vivo in mammals. Hammock and others noted in the review have characterized the principal catalytic amino acids in human EH and the chemical mechanism of action, a two-step mechanism involving formation of a hydroxyl-alkyl-enzyme intermediate. A second, rate-limiting step involves enzyme-activated water and hydrolysis of the hydroxyl-alkyl-enzyme intermediate, which regenerates the enzyme and releases the diol (Morisseau and Hammock 2005).

Comparative in vitro studies among mouse, rat, and human liver and lung tissues using butadiene indicated inverse initial rate (V_{max}/K_m) relationships among species between CYP450 formation of butadiene monoepoxide and its EH-mediated hydration to butadiene diol (Csanady et al. 1992). The mouse formed butadiene monoepoxide most rapidly, yet had the least capacity for direct EH-mediated hydration of the three species. Another comparison among species assessed activation by initial oxidation rates versus detoxification as the sum of initial rates for EH-mediated hydration and GSH conjugation (Bond et al. 1996). These studies indicated activation/

detoxification ratios for epoxybutene and diepoxybutane were highest in mouse, intermediate in rat, and lowest for humans (Bond and Medinsky 2001). These assessments are consistent with known greater carcinogenicity in mice than rats (Himmelstein et al. 1997). From the experimental evidence, the hypothesis has emerged that relative carcinogenicity of butadiene in these species depends on balance of activation versus detoxification; carcinogenicity is increased by greater oxidation of butadiene and metabolites to electrophilic epoxides and is decreased by destruction of these epoxides via EH-mediated hydration and GST-facilitated GSH conjugation.

It would seem reasonable that the hypothesis can be extended to humans as being an even less sensitive species due to favorable balance of epoxide hydration activity, as has been stated (Bond et al. 1996; Bond and Medinsky 2001; Boogaard and Bond 1996). Toward that end, Jackson et al. (2000a) have summarized comparative in vitro studies conducted with tissues from mice, rats, and humans which indicate that relative rates of oxidation of butadiene to EB are mice > rats ≈ humans, whereas the order of oxidation rates for EB to DEB is mice > rats ≈ humans. The relative extent of EH-catalyzed hydration of EB or DEB is human > rats > mice, while GST-mediated conjugation of EB was mice ≈ rats > humans. Jackson et al. (2000a) suggested that EBdiol potentially is the most significant metabolite in humans and should be studied further.

Several research groups have constructed mathematical models in attempts to describe the flux of butadiene from inhalation through the metabolic routes to excretion (Csanady et al. 1996; Filser et al. 1993; Johanson and Filser 1993, 1996; Kohn 1997; Kohn and Melnick 1993, 1996, 2000; Leavens and Bond 1996; Medinsky et al. 1994; Melnick and Kohn 2000; Sweeney et al. 1996, 1997, 2001). These computer programs, called physiologically based pharmacokinetic (PBPK) models, rely on physiological, chemical, and biochemical constants that are consistent with specific chemical properties, physiological capacities, and biochemical limitations of species including humans. These models serve as dynamic hypotheses that predict chemical concentrations in tissues during absorption, metabolism, and disposition. Through use of representative physiological parameters and species-specific metabolic constants, these PBPK models offer unique power to extrapolate predictions of tissue toxicant concentrations across species (Krewski et al. 1994). PBPK models have offered insights into disposition of butadiene and have provided kinetic constants for formation of butadiene metabolites among species. Such models have contributed significant insight and confirmation of the relative significance of DEB in the particular sensitivity of mice to butadiene (Csanady et al. 1996). However, these PBPK models have not yet been developed with sufficient capacities to accurately predict formation and elimination of the potentially significant metabolite, EBdiol, in humans (Jackson et al. 2000a).

Perhaps incorporation of biochemical constants estimated from more physiologically complete in vitro systems, such as isolated hepatocytes (Kemper et al. 2001), may offer the necessary model sophistication.

D. Biomarkers

Biomolecules that bear imprints from chemical electrophilic transformations of butadiene can be used to assess exposure and, with sufficient validation, potentials for toxicity in animals and man. These altered biomolecules have been termed "biomarkers," measurable internal indicators of change at the molecular or cellular levels (Bennett and Waters 2000) that can signal key events between exposure and adverse health consequences. Simple biomarkers of exposure to butadiene include water-soluble metabolites of butadiene in urine. Urinary metabolites of butadiene that have been confirmed in mammals include 1,2-dihydroxy-4-(*N*-acetylcysteinyl)-butane (M-I) and the regioisomeric mixture of 1-hydroxy-2-(*N*-acetylcysteinyl)-3-butene and 1-(*N*-acetylcysteinyl)-2-hydroxy-3-butene (M-II) (Albertini et al. 2003; Bechtold et al. 1994; Blair et al. 2000; Sabourin et al. 1992; Sapkota et al. 2006). The sum of these excretion products comprised 50%–90% of [^{14}C]-labeled urinary metabolites when [^{14}C]-butadiene was administered at 8,000 ppm to mice, rats, hamsters, and monkeys (Sabourin et al. 1992). Metabolite M-I, but not M-II, was detected in human urine (Bechtold et al. 1994), with higher levels noted in workers with higher nominal exposure. The detailed mechanism of formation of M-I has not been determined, although a feasible mechanism has been proposed involving oxidation of 1,2-dihydroxybutene to a ketone intermediate, followed by Michael addition with glutathione, and then subsequent reduction. The proposed mechanism is consistent with loss of one deuterium from $^2H_6$-labeled butadiene during formation of urinary metabolite M-I (Sabourin et al. 1992).

Sapkota et al. (2006) measured ambient air concentrations (near 1 ppb) and these urinary biomarkers from tollbooth workers and compared these with urinary biomarkers excreted from laboratory personnel during weekdays or on weekends. Ambient air mean concentrations ranged from 0.4 to 1.08 ppb, whereas the M-I metabolite mean range was 258–378 ng/mL and M-II mean range was 6.0–9.7 ng/mL. Tollbooth workers had highest exposure levels and urinary biomarkers compared to the other two groups, but differences were not statistically significant among groups. Levels of the urinary metabolites were highly correlated within samples, indicating precision of the assays.

Beyond simple biomarkers of exposure, such as the water-soluble metabolites of butadiene in urine (Osterman-Golkar et al. 1991), there is keen interest in developing biomarkers of effect that are correlated or as directly related as possible with carcinogenic effects of butadiene (Osterman-Golkar and Bond 1996). Such biomarkers of the toxic covalent reactions with cellular nucleophiles include fragments that appear in urine from butadiene metabolite–DNA adducts (Elfarra et al. 1995; Fustinoni et al. 2002; Osterman-Golkar and Bond 1996; van Sittert et al. 2000), butadiene metabolite–DNA adducts taken from accessible cells or tissues (Fred et al. 2004b; Koc et al. 1999; Osterman-Golkar et al. 1991, 1998; Powley et al. 2005; Shuker 2002; Zhao et al. 2000, 2001), or adducts derived from butadiene metabolites bound to hemoglobin (Elfarra et al. 2001b; Osterman-Golkar and Bond 1996; Perez et al. 1997; Powley et al. 2005; Richardson et al. 1996; van Sittert et al. 2000). N1-(2,3,4-trihydroxybutyl)adenine adducts induced by 1,3-butadiene (BD) were analyzed from lymphocytes of 15 butadiene-exposed workers and 11 control workers by the ^{32}P-postlabeling technique using high-pressure liquid chromatography (HPLC) with radioactivity detection. Difference in adduct levels between the exposed workers (4.5 adducts/10^9 nucleotides) and the controls (0.8 adducts/10^9 nucleotides) was statistically significant, suggesting that N1-THB–adenine adducts may be used for monitoring of human butadiene exposure (Perez et al. 1997; Zhao et al. 2000).

Even though not directly related to mutagenic action, covalent adducts of butadiene metabolites with hemoglobin protein serve as surrogates for DNA adducts and can integrate exposure to reactive electrophilic metabolites over periods up to the life of the red cell (Tornqvist and Landin 1995). Sun et al. (1989) appear to have conducted the first studies that indicated formation of hemoglobin adducts after administration of butadiene. Using ^{14}C-labeled butadiene, they noted that hemoglobin adduct levels were linearly related to butadiene doses of up to 100 μmol/kg for mice or rats. The butadiene-derived hemoglobin adducts in blood had lifetimes of about 24 and 65 d in mice and rats, corresponding to lifetimes for their red blood cells, respectively. These and more recent studies in exposed rodents (Boysen et al. 2004; Fred et al. 2004b; Osterman-Golkar et al. 1991, 1998; Perez et al. 1997; Powley et al. 2005; Richardson et al. 1996; Swenberg et al. 2000) and in human workers (Albertini et al. 2003; Hayes et al. 2000; Osterman-Golkar et al. 1996; Osterman-Golkar and Bond 1996; Tornqvist and Landin 1995; van Sittert et al. 2000) indicated that measurements of hemoglobin adducts formed in vivo from butadiene are useful as internal biomarkers of exposure to butadiene metabolites.

A detailed analysis of biomarker use in an extensive epidemiological study (Albertini et al. 2003) explored the advantages and disadvantages for use of two valine adducts, *N*-(2-hydroxybutenyl)-valine (HBVal) and *N*-(1,2,3-trihydroxybutyl)-valine (THBVal) (see Fig. 2), as biomarkers of exposure. Levels of THBVal were noted to be as much as 600 times higher than HBVal within the same exposure group. Rather than providing increased sensitivity as a biomarker of butadiene exposure, the higher levels of THBVal were attributed to formation from unknown endogenous sources. Therefore, HBVal may provide a better signal-to-noise ratio for detection of low levels of butadiene exposure (Albertini et al. 2003).

Moll and Elfarra (1999) studied in detail adducts formed from the monoepoxide of butadiene, EB, with the N-terminal valine of rats and mice in vitro. Analyses were conducted by an adaptation of the classic modified Edman method (Tornqvist et al. 1986, 2002). These studies described chemical adduct formation between the terminal carbon of the epoxide with the nitrogen of valine. Second-order rate constants for that reaction were 4.6- and 26 fold higher in mice and rats, respectively, than reaction with the other (C-2) carbon of EB. This apparent preference for the terminal carbon is reflected in Fig. 2, in which other potential hemoglobin adduct regioisomers are not shown for sake of clarity. A more comprehensive structural analysis of hemoglobin adducts from EB using a proteomics approach has been presented (Moll et al. 2000). This study, which relied on digestion of globin with trypsin or hydrochloric acid, indicated EB-globin-binding sites that included the α-amino group of valine, the ε-amino group of lysine, the imidazole nitrogen of histidine, and other nucleophilic sites, including serine and methionine. Six sites on α-globin and two sites on β-globin were noted as being most highly reactive. Swenberg et al. (2001) reviewed knowledge derived from DNA and hemoglobin adducts from butadiene and discussed their utility for reducing the sources uncertainty that plagues epidemiological risk assessment of exposure to butadiene.

Albertini et al. (2003) noted that no method had been available to measure specific hemoglobin adducts from DEB, as the bis-linkage of the diepoxide valine adduct and formed tertiary amino group is not suitable for the modified Edman reaction, which requires a replaceable hydrogen. Reaction of DEB with the valine primary amino group gives a cyclic adduct with a closed ring pyrrolidine-like compound, *N,N*-(2,3-dihydroxy-1,4-butadiyl)-valine (Pyr-Val) (Rydberg et al. 1996). Swenberg and coworkers (Boysen et al. 2004) developed means to measure this Pyr-Val adduct from DEB using trypsin cleavage of the globin, followed by immunoaffinity column purification of adducted peptides, and capillary liquid chromatography with electrospray ionization, tandem mass spectrometry, and selected reaction monitoring. The significance of this advance is to distinguish adducts that form only from DEB, rather that via EB or EBdiol. Using this methodology, these authors noted that mice were much more efficient in forming DEB than rats, which is consistent with the greater carcinogenicity of butadiene in mice and greater genotoxicity of DEB. Fred et al. (2005) used the Pyr-Val hemoglobin adduct biomarker to examine systemic dose of DEB formed following treatment of rats and mice with EB, as calculated from adduct levels determined in vivo and the rate constant of DEB reaction with hemoglobin determined in vitro. This study noted, after normalization to administered monoepoxide dose, that the calculated amount of DEB formed in mice was twice as high as the corresponding diepoxide formed from isoprene epoxide. In similar rat studies, the Pyr-Val adduct was below limits of quantification of the method at all levels of administered EB. The latter result is consistent with the hypothesis regarding increased detoxification capability of rats versus increased oxidative action in mice.

E. Genotoxicity

Studies on the genetic toxic effects of butadiene exposure have been reviewed (Himmelstein et al. 1997; Jackson et al. 2000a; Jacobson-Kram and Rosenthal 1995). As already noted, butadiene is indirectly toxic to genetic material, in that these actions occur as a result of oxidative metabolites. Himmelstein et al. (1997) notes that EB and DEB differ in their potential reactions with DNA, as DEB exhibits two reactive epoxides and can serve to cross-link between nucleophilic sites in DNA or proteins (Cochrane and Skopek 1994a; Lawley and Brookes 1967).

Studies have noted a variety of genotoxic effects that result from reactive metabolites following butadiene exposure. Genotoxic effects beyond DNA alkylation involved cytogenetic effects including induction of micronuclei in developing erythrocytes and sister chromatid exchange in cytogenetic studies of bone marrow cells from mice exposed to butadiene (Tice et al. 1987). Groups of male B6C3F1 mice were exposed to butadiene at 6.25, 62.5, and 625 ppm for 10 d. In bone marrow, this exposure induced a significant increase in the frequency of chromosomal aberrations, a significant elevation in the frequency of sister chromatid exchanges, a significant lengthening of average generation time, and a significant depression in the mitotic index. Noted in peripheral blood were significant increases in the proportion of circulating polychromatic erythrocytes and in the levels of micronucleated polychromatic erythrocytes and micronucleated normochromatic erythrocytes. The most sensitive indicator of genotoxic damage was the frequency of sister chromatid exchange, which was significant at 6.25 ppm (Tice et al. 1987). Fred et al. (2005) showed the increase in frequencies of polychromatic erythrocyte micronuclei in mice were approximately linearly correlated to in vivo formed doses of DEB, as determined through use of the Pyr-Val hemoglobin adduct following administration of EB (see above). In rats treated with EB, no significant increase in polychromatic erythrocyte micronuclei was detected, which is consistent with lack of quantifiable levels of Pyr-Val hemoglobin adducts observed in the study (Fred et al. 2005).

Butadiene was mutagenic in vivo at the hypoxanthine guanine phosphoribosyltransferase (hprt) locus of splenic T cells taken from B6C3F1 mice exposed to 625 ppm for 2 wk (Cochrane and Skopek 1994b). DNA sequencing revealed that about half of the mutations induced in vivo by butadiene, EB, and DEB were frameshift mutations, while remaining mutations produced by butadiene, EB, and DEB were transition and transversion mutations at both AT and GC base pairs.

Recio et al. (2000) examined mutagenicity and the mutational spectra of EB and DEB in human and rodent cells. In human TK6 lymphoblastoid cells, EB treatment in vitro increased the frequency of GC to AT transitions and AT to TA transversions at the hprt locus in the TK6 cells, and sequences were reported. DEB treatment caused AT to TA transversions and increased frequency of base deletions. The authors concluded that at the hprt locus in human cells, EB was genotoxic mainly through point mutations, while DEB caused deleterious point mutations and partial deletions. The study noted a particular role for purine adducts in induction of mutation in human cells and indicated that guanine and adenine adducts should serve as biomarkers of exposure in humans. In vitro and vivo rat studies from this group (Recio et al. 2000) also examined lacI locus mutant frequencies from the two butadiene epoxide metabolites and noted increased frequencies of three types of base substitution mutations, which included GC to AT transitions, GC to TA transversions, and AT to TA transversions. The detailed sequence data reported should provide a basis for future molecular studies.

Observations of mutations of the endogenous hprt gene in animals were extended in a pilot study examining genotoxicity in human peripheral blood lymphocytes from workers in a butadiene production plant. That study (Ward et al. 1994) noted elevated mutant frequencies of the hprt gene in T cells from workers with higher exposures, and these were highly correlated with urinary concentrations of the urinary metabolite M-I. Unexposed cohorts lacked the elevated mutation rate and urinary M-I metabolite levels.

Other studies conducted to gain understanding of the types of mutations that arise following butadiene exposure have included known cellular signal transduction pathways. *ras* protein is a GTP-binding protein that serves as a signal transducer in many types of cells (Satoh and Kaziro 1992). This signaling pathway is involved in many activities of cellular growth and maintenance, including differentiation, transformation, and proliferation (Evans et al. 1991; Satoh and Kaziro 1992). Mutational activation of *ras* oncogenes, leading to lack of controlled differentiation and proliferation, has been noted among the earliest detectable changes in mammary neoplastic cells following administration of chemical carcinogens (Kumar et al. 1990; Sukumar 1990). Analysis of these proto-oncogenes in liver and lung tumors from butadiene-exposed mice (Goodrow et al. 1990; Sills et al. 1999a) have noted activation of K-*ras* genes via G to C transversion at codon 13, which was not observed in spontaneous tumors from unexposed animals. Zhuang et al. (2002) have suggested that the commonly observed G to C base substitution in *ras* genes may result from direct interaction of reactive epoxides of butadiene with the *ras* genes.

Alterations of tumor suppressor genes, including loss of heterozygosity in the p53 locus in lymphomas and lung tumors, have been observed following exposure to butadiene (Wiseman et al. 1994; Zhuang et al. 1997). Mutations in exons 5 to 8 of the p53 gene were associated with hemangiosarcomas in B6C3F1 mice treated with 200 and 625 ppm butadiene

(Hong et al. 2000), and similar inactivation of p53 was observed in butadiene-induced mammary adenocarcinomas (Zhuang et al. 2002). In contrast to specific base substitutions noted with *ras* genes, a variety of p53 gene alterations were observed, which may reflect the different number of sites that can degrade function of p53 protein, versus more specific activation of *ras* function. The p53 protein plays a major role in regulation of the cell cycle, determining if cells continue the cycle of cell division or enter senescence pathways leading to apoptotic degradation (Ben Porath and Weinberg 2005; Hirabayashi 2005). Additionally p53 protein appears to modulate DNA repair mechanisms, as a direct link has been reported between p53 protein and nucleotide excision repair mechanisms (van Steeg 2001). Damage to this pathway prevents timely maintenance of carcinogen-damaged genetic structures within cells, and leads to additional mutations in addition to accelerated cell division.

Additional missense mutations have been noted within genes encoding the Wnt/β-catenin cellular signaling pathway in mammary adenocarcinomas from mice following butadiene exposure (Zhuang et al. 2002). The Wnt signaling cascade, which involves β-catenin as a critical signal component, is important for renewal of cells having rapid turnover, such as intestinal epithelial stem cells, epidermal stem cells associated with hair follicles, and hemopoietic stem cells (Reya and Clevers 2005). Point mutation in codons 33, 34, and 42 of the β-catenin gene alter turnover of this component and signal levels; this abnormally stimulates target genes, and is considered a critical event in malignant transformation (Reya and Clevers 2005; Zhuang et al. 2002).

Butadiene exposure results in formation of specific DNA adducts by way of nucleophilic substitution reactions with butadiene epoxide metabolites (Blair et al. 2000; Booth et al. 2004; Henderson 2001; Jackson et al. 2000a; Osterman-Golkar and Bond 1996). Major DNA adducts to the purine bases include hydroxybutenyl- and trihydroxybutanyl-substitution (THB) at the N7-sites of guanine, and at the N1-, N3-, and N^6-positions of adenine (Swenberg et al. 2001; Tretyakova et al. 1998). The N1- and N3-adenine adducts formed from EB have been noted to be unstable and lead to N^6-adenine adducts by rearrangement and to N1-inosine adducts by deamination (Rodriguez et al. 2001). The N1-inosine preferentially pairs in double-strand DNA with cytosine. When replicated, the deaminated base will give rise to an A to G point mutation at high frequency (Rodriguez et al. 2001).

Dosimetry studies of N7-guanine adduct formation by EB, DEB, and EBdiol in liver, lung, and kidney from butadiene-treated B6C3F1 mice and F344 rats were able to estimate proportions of N7-(2,3,4-trihydroxybutyl)guanine (THB-guanine) adducts that resulted from DEB and EBdiol (Koc et al. 1999). These authors and others (Boogaard et al. 2004; Booth et al. 2004; Koivisto et al. 1999), have concluded that most of THB-guanine is formed from EBdiol. One study (Koc et al. 1999) noted the THB-guanine adducts to be most abundant, and that the levels reached a plateau in rats in middose range (20, 62.5, 625 ppm), implying saturation of activation, but not in mice given the same inhalation exposure. A similar study by Oe et al. (1999) noted higher levels of THB-guanine adducts in liver tissue of mice than in rats exposed at high levels (1,250 ppm, for 2 wk), and observed that THB-guanine adducts persisted somewhat longer in liver of mice than in rats. Although the N7-guanosine adducts are by far the most abundant, their significance has been questioned because the modification at N7 renders the glycosyl bond unstable and susceptible to spontaneous depurination (Carmical et al. 2000b).

The inherent instability of N7-guanosine adducts precludes accurate determination of mutagenic potential using site-specific modified deoxynucleotides. In these sophisticated studies with stable adducts, nucleoside adducts were formed with stereoisomers of EB and EBdiol, which then were incorporated into nucleotides. These adducted oligodeoxynucleotides were ligated into shuttle vectors and incorporated transgenically into *Escherichia coli*, where mutagenic frequencies are examined in the N-*ras* 12 codon. Initial studies with N^6-adenine adducts of EB were nonmutagenic, but N2-guainine adducts from different enantiomers of EBdiol were stereospecific in the mutations fixed. Mutations generated by adducts from the R,R EBdiol enantiomer misincorporated A to G exclusively, while adducts of the S,S EBdiol enantiomer yielded exclusively A to C mutations (Carmical et al. 2000a).

More-recent studies have examined perturbations in the helical structure and hydrogen bonding of double-stranded DNA, which involve impact of trihydroxybutyl adducts to N^6-adenine in synthetic oligimers that simulate codon 61 of the human N-*ras* ongogene (Merritt et al. 2004, 2005; Scholdberg et al. 2004, 2005). Mispairings of bases and effects on DNA geometry have been categorized, with suggestions that DNA polymerases read through the mispairings, resulting in significant mutagenicity.

Powley et al. (2003) have described novel substituted 1,*N*2-propanodeoxyguanosine adducts formed by Michael addition of the nonepoxide butadiene metabolite, 1-hydroxybut-3-en-2-one, otherwise known as hydroxymethylvinyl ketone (HMVK). The significance of these adducts, which include diastereomeric and positional isomers, is not yet known, but they present an alternative hypothesis to the theory that carcinogenicity arises as a result of the quantities of adducts formed from epoxide metabolites.

F. Human Epidemiology

The International Agency for Research on Cancer (IARC), part of the World Health Organization, periodically reviews scientific data published worldwide regarding epidemiological and experimental cancer studies. IARC summarizes findings with respect to cancer causation by chemicals and has provided a comprehensive review of butadiene, including production data, metabolism, toxicity, and epidemiological studies (IARC 1999a). IARC classified butadiene in this most recent overall evaluation in *Group 2A*, as *probably carcinogenic to humans*. This overall evaluation is derived from determinations that there is *limited evidence* in humans, but there is *sufficient evidence* in experimental animals, for the carcinogenicity of 1,3-butadiene. The likelihood, short of certainty, of human butadiene carcinogenicity was derived primarily from one large, well-conducted study of workers in the styrene-butadiene rubber industry (Delzell et al. 1995, 1996) that indicated excess leukemia in more highly exposed workers relative to those with lower exposure (Macaluso et al. 1996; Sathiakumar et al. 1998). As noted (IARC 1999a), "the evidence from this study strongly suggests a (cancer) hazard, but the body of evidence does not provide an opportunity to assess the consistency of results among two or more studies of adequate statistical power." A more recent follow-up to the previous study of Sathiakumar et al. (1998) (Graff et al. 2005) provided weight to the positive association between cumulative butadiene exposure and deaths from all forms of leukemia. Chronic mylogenous leukemia and chronic lymphocytic leukemia indicating highest relative rates for death, 7.2- and 3.9 fold at highest exposure levels, respectively.

A risk characterization for butadiene has been published by the Environmental Health Directorate of Canada (Hughes et al. 2003), which succinctly summarized epidemiological data and analyzed exposure-response functions from the studies of Delzell et al. (1995, 1996). This study concluded that "exposure to 1,3-butadiene in the occupational environment has been associated with the induction of leukemia; there is also some limited evidence that 1,3-butadiene is genotoxic in exposed workers. Therefore, in view of the weight of evidence of available epidemiological and toxicological data, 1,3-butadiene is considered highly likely to be carcinogenic, and likely to be genotoxic, in humans" (Hughes et al. 2003).

The U.S. EPA has classified butadiene as a human carcinogen (USEPA and IRIS 2002), consistent with findings of the National Toxicology Program that butadiene is *known to be a human carcinogen* (NTP 2005a). The latter reference summarizes and references significant toxicological studies leading to this determination. Findings of excess hematopoietic cancers from epidemiological studies of plant workers were determined to be consistent with causal association with butadiene exposure. This conclusion was justified based on causality criteria of temporality, strength of association, specificity, biological gradient, and consistency. Mechanistic similarity of metabolic pathways among mammals was said to fulfill the criterion of biological plausibility. Based on these considerations, human epidemiological data were deemed sufficient for the classification of butadiene as a human carcinogen. Again, the primary epidemiological study driving these conclusions is that of Delzell et al. (1995, 1996), which evaluated the mortality of 15,649 men employed at any of eight North American styrene-butadiene rubber plants for at least 1 year. A greater than statistically expected number of deaths due to leukemia occurred in subjects who had ever worked hourly more than 10 yr and were 20 yr beyond hiring (standard mortality ratio, SMR = 1.43 fold over expected mortality), subjects working in polymerization (SMR = 2.51), in maintenance labor (SMR = 2.65), and in laboratories (SMR = 4.31). All 95% confidence intervals exceeded the expected SMR = 1.00 (Delzell et al. 1996).

G. Risk Assessment

A quantitative cancer risk potency was determined from epidemiological data (Delzell et al. 1995, 1996) by the U.S. EPA (USEPA 2002), by determination of a dose–response function based on linear extrapolation from 1% leukemia death incidence (modeled from the epidemiology data) to zero exposure. The slope of this function was arbitrarily doubled by EPA based on the increased sensitivity of mice and the supposition that the human "epidemiology-based estimate may underestimate total cancer risk from 1,3-butadiene exposure in the general population . . . , resulting in a lifetime excess cancer unit risk estimate of 8×10^{-2}/ppm (USEPA 2002)." This value translates into 8×10^{-5}/ppb or 3.6×10^{-5} µg/m^3. Using these slope estimates, the level of lifetime chronic butadiene exposure giving an estimated 1 in 1 million risk is 22 ng/m^3 (Sapkota et al. 2006), or 9.9 ppt butadiene. As noted earlier, many areas of the country exhibit ambient air concentrations substantially above this value. It is important to realize that estimation of such risk estimates is replete with uncertainty (Krewski et al. 1999) and that valid scientific objections have been raised against the default linear extrapolation to zero exposure for risk estimation (Waddell 2002, 2003a,b,c, 2005). The answer to whether this estimate of potential risk forecasts real cases of leukemia in the population awaits future research with more-refined methodology.

III. Chloroprene

A. Chemical Information and Use

Chloroprene (CAS No. 126-99-8) (see Fig.1) is a colorless synthetic liquid with a pungent, ether-like odor. It is soluble in ethanol and diethyl ether, miscible with acetone and benzene, and slightly soluble in water. Other physical properties are given in Table 1. Currently, chloroprene is produced by chlorination of 1,3-butadiene to 3,4-dichloro-1-butene, with subsequent caustic dehydrochlorination to 2-chloro-1-3-butadiene (Lynch 2001c). Production worldwide has been indicated to be of the order of 100,000 t annually (NTP 2005b). For economic reasons, this synthetic route largely has replaced an older method used before 1960 that involved addition of two acetylene molecules to form vinylacetylene, followed by hydrochlorination to chloroprene. Chloroprene monomer can be polymerized with near-100% yield, but is extremely reactive. Multiple undesirable reactions, including spontaneous polymerization, dimerization, oxidization, epoxidation and nitration, can occur (Lynch 2001c). As a consequence chloroprene must be stored cold under nitrogen in the presence of reaction inhibitors. Because of this reactivity chloroprene is not generally available from chemical supply houses.

Exposure potential for workers is less in production of chloroprene monomer than in polymerization facilities, as the latter are batch processes that involve more maintenance and are usually conducted in closed buildings (Lynch 2001c). Data on chloroprene personnel monitoring during 1976–1996 indicate that polymer workers were exposed to levels ranging from 1 to 8 ppm. Monomer production worker exposure was in the range of 1 to 2 ppm or below during that period (Lynch 2001b).

Chloroprene is not known to occur naturally. There are less than a dozen sites worldwide where chloroprene is manufactured, and most of these are listed by Lynch (2001c). In contrast to butadiene, production is more limited, and therefore so is environmental release. U.S. EPA Toxic Release Inventory data for 2003 estimate total environmental release of chloroprene equivalent to 382 t; 89% occurred as point-source releases into air, 8.3% as fugitive air emissions, and 2.5% was disposed by underground injection (USEPA 2005d).

Given the limited number of production sites and lack of natural occurrence, data regarding chloroprene concentrations in ambient air are scarce. Such data exist from air monitoring in western Louisville, Kentucky, for an area known as Rubbertown that includes 11 petrochemical plants. Here, chloroprene is polymerized into polychloroprene. Air monitoring data from samples taken by EPA procedure (USEPA 2000) is posted on the website of the West Jefferson County Community Task Force (WJCCTF 2006), a community-based organization devoted to assessment and resolution of pollution issues in the area. These data, taken over the period 2000–2006, indicate chloroprene ambient air concentrations that exceed 1 ppb at several sites, including a public park and outside an elementary school. Maximal levels were noted at a monitoring site near the plant, with ambient levels spiking at levels greater than 10 ppb on more than 10 during over the monitoring period.

B. Toxicity

Toxicity studies for chloroprene have been reviewed (Valentine and Himmelstein 2001). There have been questions regarding purity of chloroprene used in some early studies, as the reactivity of chloroprene with itself or with air was not widely understood or considered. Acute lethal human exposure to chloroprene is associated with nervous system depression, pulmonary edema, narcosis, and respiratory arrest (Nystrom 1948). Chloroprene is acutely lethal in rats exposed at levels of 2,300 ppm for 4 hr, with primary toxic effects including pulmonary hemorrhage and edema and hepatic necrosis. A threshold for chloroprene-induced acute hepatotoxicity in rats by 4-hr exposure has been estimated as between 106 and 180 ppm (Plugge and Jaeger 1979). Inhalation studies in rats for 2 and 4 wk had early deaths at levels greater than 161 ppm, with olfactory epithelial degeneration, centrilobular hepatic necrosis, and decreased red blood cell counts (Clary et al. 1978). Subchronic 90-d inhalation in F344 rats produced similar effects, with observations including metaplasia of the olfactory epithelium and loss of nonprotein sulfhydryl content of lungs and liver at 80 ppm exposure. A 90-d subchronic inhalation study in B6C3F1 mice noted deaths at 200 ppm, and hyperplasia of the forestomach epithelium (Melnick et al. 1996).

Early reproductive and developmental studies suggested that chloroprene exhibited embryotoxic effects, but these studies did not indicate purity or means of generation of chloroprene vapor for inhalation studies, and have been discounted (Valentine and Himmelstein 2001). Mutagenicity studies, as reviewed by Valentine and Himmelstein (2001), using the Ames *Salmonella* test (Ames et al. 1973b; Ames 1973) were unremarkable with pure chloroprene alone but proved positive with addition of liver homogenate supernatant for compound metabolic activation (Ames et al. 1973a). Subsequent positive mutagenicity tests with (1-chloroethenyl)oxirane, a primary oxidative metabolite of chloroprene, indicated mutagenicity comparable to that of epoxybutene (Himmelstein et al. 2001b).

Two-year chloroprene inhalation studies were conducted by the National Toxicology Program in F344/N rats and B6C3F1 mice (Melnick et al. 1999a; NTP 1998). Chloroprene was carcinogenic in F344/N rats, with observations of papilloma or carcinoma of oral cavities, adenoma or carcinoma of thyroid follicular cells, kidney renal tubules, and lung alveolar or bronchiolar cells, and fibroadenoma of mammary gland (Melnick and Sills 2001). These effects were noted at inhalation concentrations of 12.8 ppm for renal tubule cancers, or at 32 ppm for most other tumors. Similar tumors were observed in B6C3F1 mice at the low inhalation group (12.8 ppm), with additional sites of adenomas or carcinomas including the Harderian gland, skin, mesentery, liver, and circulatory system (hemangioma and hemangiosarcoma) (Melnick and Sills 2001). Dose–response calculations indicated that the most sensitive response was induction of lung neoplasms in female mice, which yielded ED_{10} (calculated 10% response) values of 0.3 ppm. The latter paper summarized and compared tumor induction in mice and rats exposed to chloroprene, butadiene, isoprene, or ethylene oxide. All these compounds are believed carcinogenic due to epoxide metabolites (Melnick and Sills 2001). Thus, chloroprene exhibits toxicity substantially similar to that previously detailed with butadiene.

C. Metabolism and Kinetics

The first indication that metabolism was involved in producing toxic metabolite(s) of chloroprene came from early studies of its mutagenicity in

Salmonella typhimurium strains (Bartsch et al. 1979). Incubation of chloroprene with hepatic microsomes produced a volatile metabolite that was proposed to be an epoxide based on analogy with other halogenated unsaturated compounds. Follow-up studies in isolated hepatocytes and rats treated orally with large doses of chloroprene indicated that chloroprene caused depletion of GSH with excretion of urinary thioethers, and that oxidative metabolism was prerequisite. It was noted that chloroprene does not react readily with GSH nonenzymatically, or in presence of GSH and glutathion S-transferases. Much more recent work (Himmelstein et al. 2001a) detected (1-chloroethenyl)oxirane (1CEO) in an organic extract of an incubation of chloroprene with rat liver microsomes, thereby confirming the previous hypothesis that an epoxide metabolite was formed. Using the enzyme inhibitor 4-methyl pyrazole, the latter study strongly suggested that 1CEO was produced by CYP 2E1-mediated oxidation of chloroprene. Further studies with synthetic 1CEO noted that its hydrolysis was inhibited with 1,1,1-trichloropropene oxide, a competitive inhibitor of epoxide hydrolase (EH). These findings extended understanding of the similarities between chloroprene and butadiene with respect to metabolic activation and detoxification. Hypotheses for testing of chloroprene have been based on inference from studies of butadiene, although less consideration has been given to critical evaluation of differences in disposition caused by molecular differences.

A significant series of papers (Cottrell et al. 2001; Munter et al. 2002, 2003) has proposed details of metabolism of chloroprene based on in vitro metabolic experiments, extensive synthesis of metabolites, and analytical determinations by chromatography, mass spectrometry, and nuclear magnetic resonance spectroscopy. This work examined chemical and metabolic reactions involving chloroprene bioactivation by CYP450 oxidation to 1CEO, through detoxification with EH and GSH, and has indicated potential adducts of 1CEO with DNA in vitro. A summation of the chemical and metabolic pathways is shown in Fig. 3, which includes reactions that have been demonstrated in vivo or proposed from in vitro studies. Two epoxide metabolites were proposed, which would seem analogous to the epoxidation of butadiene. From kinetic studies of rodent tissue metabolism (Himmelstein et al. 2004b), the amount of the major electrophilic metabolite, 1CEO, has been estimated as 2%–5% of dose. As is evident from Fig. 1, the fundamental difference between chloroprene and butadiene is substitution of a chlorine atom for a hydrogen atom at the 2-carbon. Alternative chemical epoxidation of the double bond to 2-chloro-2-ethenyloxirane (2CEO) produced an epoxide that is significantly less stable than 1CEO in aqueous media (Cottrell et al. 2001). As noted by these authors, the difference in electronegativity of the chlorine substitution leads to rearrangements and production of chloroaldehydes and ketones. Therefore no stable diepoxide analogous to the extremely mutagenic, bifunctional butadiene metabolite DEB has been detected for chloroprene in biological systems (Munter et al. 2003).

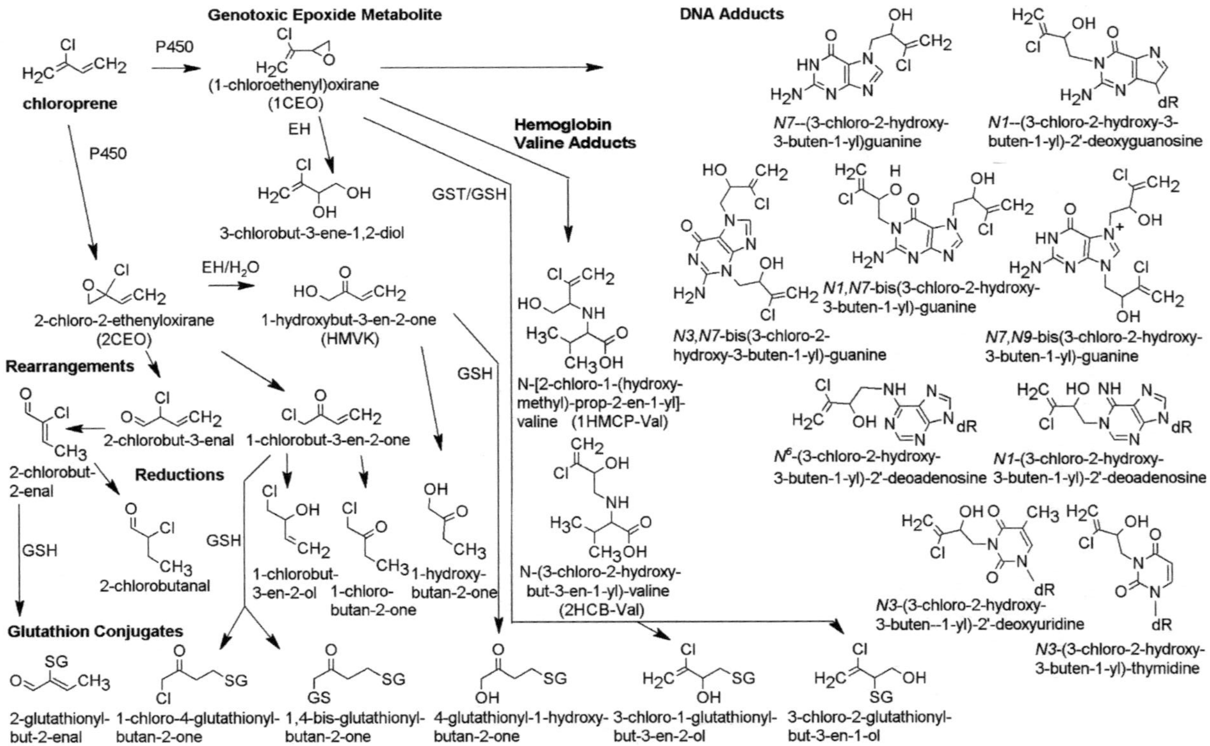

Fig. 3. Metabolic scheme for chloroprene. *Note*: 1-hydroxy- and 2-hydroxy adduct regiosiomer pairs may exist from electrophile–nucleophile reactions that are not shown for simplicity.

The roles in the toxicity of chloroprene of the chloro-aldehydes and ketones (see Fig. 3) formed by rearrangements are yet unclear. It is possible, however, that this pathway of rearrangements directed by the chlorine substitution has major biological significance due to apparent lack of formation of a stable diepoxide. Note that a metabolite in common with butadiene, 1-hydroxybut-3-en-2-one (HMVK), may be formed by hydrolytic cleavage and chlorine loss from 2CEO (Cottrell et al. 2001).

As with butadiene, epoxide hydrolase plays an important role in detoxification. This microsomal enzyme hydrates the electrophilic epoxide moiety of 1CEO to give much less toxic 3-chlorobut-3-en-1,2-diol (Munter et al. 2003). In vitro studies of microsomes from mouse, rat, and human indicate that the mouse forms 1CEO more rapidly than the rat or human (Munter et al. 2003), particularly in mouse lung tissue (Himmelstein et al. 2004b). As noted with butadiene, such increased production of 1CEO portends greater formation of adducts by electrophilic attack on nucleophilic sites of DNA, as shown in Fig. 3 (Munter et al. 2002).

Reaction of 1CEO with GSH formed 3-chloro-1-(*S*-glutathionyl)but-3-en-2-ol as a major product, with the regioisomer 3-chloro-2-(*S*-glutathionyl)but-3-en-1-ol as a minor conjugate in vitro (Munter et al. 2003). A number of glutathione conjugates arise from the epoxides and the α-,β-unsaturated ketone metabolites (Munter et al. 2003). While not yet shown through in vivo studies, these likely will lead to elimination of N-acetylcysteinyl-substituted urinary metabolites as observed with butadiene.

D. Biomarkers

Compared to butadiene, much less work has been accomplished with respect to biomarkers of chloroprene exposure. Hemoglobin valine adducts have been described following in vitro exposure of erythrocytes to 1CEO (Hurst and Ali 2006). No reports have been published that involve determination of biomarkers following in vivo administration, perhaps due to known issues of chloroprene stability and lack of its general availability as noted above. Based on the common HMVK metabolite (see Figs. 2 and 3), it is possible that the butadiene urinary metabolite 1,2-dihydroxy-4-(*N*-acetylcysteinyl-*S*)-butane (M-I) could be a urinary marker for chloroprene exposure.

E. Genotoxicity

Interpretation of early genotoxicity studies of chloroprene following in vitro tests has been difficult because of conflicting results in differing systems. Initial Ames tests in *Salmonella typhimurium* strains TA100 and TA1530 (Bartsch et al. 1975, 1976; IARC 1979) indicated that chloroprene was mutagenic with or without metabolic activation by induced liver homogenate enzymes. However, Zeiger et al. (1987) did not observe mutagenic activity of chloroprene in *Salmonella* strains, including TA100.

Clarification of these conflicting results was provided by a study using freshly distilled chloroprene (Westphal et al. 1994) in which no mutagenicity in *S. typhimurium* strain TA 100 was noted with purified chloroprene, whereas mutagenic effects were observed (Westphal et al. 1994) with aged chloroprene that correlated with formation of cyclic dimers of chloroprene. Tests for chloroprene mutagenicity in cultured V79 Chinese hamster cells were negative (Drevon and Kuroki 1979).

Results of in vivo mammalian genotoxicity tests similarly are in conflict, as dominant lethal mutations were induced by chloroprene in germ cells of male rats and mice, as well as chromosomal aberrations in bone marrow cells of mice (IARC 1979; Sanotskii 1976). Another study (Tice et al. 1988) presented contrasting results, as chloroprene concentrations of 80 ppm or less did not increase chromosomal aberrations, did not change the rate of erythropoiesis, and did not alter bone marrow cell proliferation. However, Tice et al. (1988) observed significant increases in the mitotic index in the bone marrow of chloroprene-treated mice.

Given these conflicting results that seem to be dependent on the type of genotoxicity assay and the questionable purity of the chloroprene exposure preparation, more detailed examination of specific mutation frequencies in *ras* proto-oncogenes was undertaken (Sills et al. 1999b) following the 2-yr inhalation chloroprene carcinogenicity studies of the National Toxicology Program (Melnick et al. 1999a). This study (Sills et al. 1999b) noted significantly increased numbers of K-*ras* mutations in chloroprene-induced mouse lung neoplasms and K- and H-*ras* mutations in the Harderian gland of B6C3F1 mice. The most frequent base change was an A to T transversion at K-*ras* codon 61, and this mutation was suggested to be important for chloroprene tumor induction.

F. Human Epidemiology

There have been a limited number of studies examining the epidemiology of exposure to chloroprene in the workplace, reviewed by Acquavella and Leonard (2001). Most early studies are not in the English literature, and may have not adequately characterized exposure. A report of epidemiological studies for excess cancer related to chloroprene exposure was published from the Research Laboratory of Hygiene Toxicology, West China University of Medical Sciences (Li et al. 1989). This study of 1,213 persons, of which 149 had chloroprene exposures for at least 25 yr, noted significantly increased cancer mortality in workers. Based on observed greater standard mortality ratios at higher exposures, the study suggested that chloroprene exposure increased incidence of liver, lung, and lymphatic cancers. One early U.S. study (Pell 1978) attempted to determine if exposure to chloroprene increased risk for lung cancer. The study involved two cohorts: one involving 270 men exposed between 1931 and 1948, and the other consisting of 1,576 men exposed between 1942 and 1957. Numbers of lung cancer deaths in these cohorts were similar to the number of expected deaths, and absence of excess lung cancer mortality in the groups with high exposure led to the conclusion that occupational chloroprene exposure did not increase risk of lung cancer.

Recent epidemiological studies of exposure to chloroprene have been conducted under sponsorship of the International Institute of Synthetic Rubber Producers. This study was international in scope, involving 12,430 workers at two U.S. sites (Louisville, Kentucky and Pontchartrain, Louisiana) and two European sites (Maydown, Northern Ireland, and Grenoble, France). The study included detailed exposure reconstruction and job-task classification for potential exposures to chloroprene monomer and vinyl chloride. Median values of worker exposure to chloroprene were 5.23, 0.028, 0.16, and 0.149 ppm for these sites, while median cumulative exposures were 18.3, 0.133, 0.084, and 1.01 ppm-years, respectively. Preliminary publications from this large international study (Leonard et al. 2006; Marsh et al. 2006a,b) have indicated little or no increased mortality risk from all causes, all cancers, or lung or liver cancer. One statistically significant association was seen between duration of exposure to chloroprene and all combined cancers in the Northern Ireland plant (Marsh et al. 2006b). Additional assessment awaits future analysis within these studies.

Regardless of these recent negative epidemiological studies, the *National Toxicology Program Report on Carcinogens*, 11th edition, has classified chloroprene as *reasonably anticipated to be a human carcinogen* (NTP 2005b) and summarizes significant toxicological results from studies of chloroprene. This classification was based on evidence of tumor formation at multiple sites in multiple species of experimental animals (NTP 1998), as well as limited data from occupational exposure (Li et al. 1989).

The evaluation for chloroprene given by IARC (1999b) is that there is *inadequate evidence* for the carcinogenicity of chloroprene in humans, but there is *sufficient evidence* for carcinogenicity of chloroprene in experimental animals. These determinations have produced the overall IARC evaluation that chloroprene is *possibly carcinogenic to humans (Group 2B)*.

G. Risk Assessment

Formal risk assessment or quantitative cancer risk potency for exposure to chloroprene has not yet been accomplished by the U.S. EPA. However, an analysis based on physiological toxicokinetic modeling of chloroprene has been published (Himmelstein et al. 2004a) that offers significant promise for diminishing uncertainty for future risk assessment. This study constructed a mathematical model based on physicochemical, physiological, and metabolic parameters for chloroprene using tissues from mouse, rat, hamster, and human. Comparisons of model predictions with experimental determinations of chloroprene vapor uptake from closed chambers enabled parameter sensitivity analysis and refinement of the model. From these analyses, the total amount of chloroprene metabolized per weight of lung tissue per day was determined to be the best measure of internal toxic dose when compared with the fraction of animals affected by lung tumors from previous cancer studies (Melnick et al. 1999b; Trochimowicz et al. 1998). The fraction of animals with tumors versus modeled total amount of chloroprene metabolized provided a dose–response curve that was in close agreement with actual experimental data whereas no such relationship existed for fraction of animals with tumors versus exposure concentrations. Furthermore, this analysis explained differences between species and strains that had been noted previously and offered a mechanistic basis for cross-species extrapolation. From these studies, estimates of the chloroprene concentrations required for 10% additional risk to humans were derived; these were 23, 98, and 109 ppm, respectively, for 24-hr, 7-d continuous exposure; 8-hr, 5-d; and 12-hr, 3-d intermittent exposures per week over a lifetime.

IV. Isoprene

A. Chemical Information and Use

Isoprene (CAS No. 78-79-5) (see Fig. 1) is a colorless volatile liquid that is unstable, oxidizable, and flammable. It is practically insoluble in water but miscible with ethanol or ether. Isoprene is obtained from petroleum cracking and is isolated as a by-product during production of ethylene (2%–5% of ethylene yield) (NTP 2005c). More than 1.5 million t isoprene are commercially produced worldwide each year, and some 95% of this is polymerized to poly(*cis*-1,4-isoprene) (Watson et al. 2001), the polymer that constitutes natural rubber (Taalman 1996). The remainder is used in production of butyl rubber (isobutene-isoprene copolymer), other polymers, and synthesis of terpenes (Taalman 1996). Other anthropogenic sources of isoprene include wood pulp production, oil fires, wood and other biomass burning, tobacco smoking, gasoline, and automobile exhaust (NTP 2005c).

In contrast to the other dienes already discussed, isoprene is a significant biosynthetic product that is produced naturally. It is a fundamental structure of isoprenoid biochemicals, which include cholesterol and other sterols, carotenoids, and vitamin A. Isoprene is produced by plants, animals (Sharkey 1996; Watson et al. 2001), and bacteria (Kuzma et al. 1995) in quantities that dwarf amounts produced synthetically. Emissions from plants, which are linked to photosynthesis, are estimated to be of the order of 70 million t annually, roughly equivalent to quantities of methane emitted globally (Sharkey 1996). Isoprene biological synthesis is believed to occur from acetyl-coenzyme A by way of mevalonic acid, which is converted to isopentenyl pyrophosphate, then to isomeric dimethylallyl pyrophosphate, and to isoprene by isoprene synthase (Silver and Fall 1995). An additional nonmevalonate pathway has recently been described as occurring in some bacteria (Kuzuyama and Seto 2003).

Plants produce isoprene during photosynthesis at rates equivalent to a few percent of the amount of carbon fixed and may use isoprene to modulate stress conditions, including temperature change, photooxidative stress, and chemical stress such as singlet oxygen radical reactions (Affek and Yakir 2002). Isoprene is abundant in human breath at concentrations in the range of 50–1,000 ppb (Karl et al. 2001), and is second in abundance only to acetone among hydrocarbons in the breath (Sharkey 1996). Humans endogenously produce isoprene at rates of the order of 0.15 micromoles/kg/hr, which amounts to about 17 mg/d for a 70-kg person (NTP 2005c). Estimates of isoprene intake from ambient air suggest that exogenous sources contribute 0.45–0.6 mg/d to this exposure, assuming an air isoprene concentration of 10 ppb (0.03 mg/m^3) and inspiration of 15–20 m^3 air/d (NTP 2005c). Breath concentrations of isoprene among humans are more variable than those of acetone and simple alcohols, and current research is examining the relationship of isoprene breath levels to cholesterol levels (Karl et al. 2001).

B. Toxicity

Toxicological studies of isoprene have been conducted because of its homology with butadiene. Similarities include oxidative metabolic pathways and mechanisms of detoxification. Evident toxicities are similar, such as mutagenicity of a diepoxide metabolite, induction of sister chromatid exchange in bone marrow, and induction of anemia. Other induced toxic effects are observed in the olfactory epithelium, testis, and forestomach of mice (Melnick and Sills 2001). Inhalation studies (Melnick et al. 1994) were conducted in male and female F344 rats and B6C3F1 mice at exposure concentrations of 0, 70, 220, 700, 2,200, and 7,000 ppm according to the paradigm of 6 hr/d, 5 d/wk for 26 wk, followed by a 26-wk period of observation. In rats treated at the highest level only, interstitial cell hyperplasia of the testis was observed in all male rats after 26 wk exposure. The only effect in rats following the 26-wk recovery period was a marginal increase in benign testicular interstitial cell tumors. In mice, isoprene induced cancers in multiple organs. Incidences of neoplastic tumors in liver, lung, forestomach, and Harderian gland were significantly increased at exposure levels of 700 ppm and higher. Noncancerous toxicities included spinal cord degeneration, partial rear limb paralysis, testicular atrophy, olfactory epithelial degeneration, forestomach epithelial hyperplasia, and macrocytic anemia. As the study was terminated after 52 wk, it was suspected that treatment and observation periods may not have been sufficient to observe late-developing tumors (Melnick and Sills 2001). Consequently, another chronic inhalation study (Placke et al. 1996) was conducted in B6C3F1 mice with exposure concentrations from 10 to 2,200 ppm for 40 or 80 wk, and mice were observed for 96 or 105 wk. This work produced exposure-related increases in tumors of liver, lung, Harderian gland, and forestomach, with a lowest observed effect level noted as 70 ppm. Toxic effects were compared to those of butadiene, but without the early onset of T-cell lymphomas. Potency of isoprene for neoplastic effects was assessed at one-tenth that of butadiene. As analyses, including calculation of the product of isoprene concentration and duration of exposure, were not adequate to predict tumor risk at any site, and extrapolation of tumor probability from high to low doses based on cumulative exposure was stated to be not justified by statistical models. These authors suggested that a nonlinear dose response and threshold effect level appeared to exist for tumor development (Placke et al. 1996).

A subsequent 2-yr isoprene inhalation study conducted in F344 rats by the National Toxicology Program (Melnick and Sills 2001; NTP 1999) demonstrated exposure-related increases in mammary gland, fibroadenomas in male and female rats, and interstitial cell adenomas of testis and renal tubule adenomas. Incidence of multiple fibroadenomas increased with increasing exposure in males and females. No increases were seen in mammary gland carcinomas. Dose–response curves were noted to be linear or supralinear in shape in the low-dose region (Melnick and Sills 2001).

C. Metabolism and Kinetics

The chemical and metabolic pathways for isoprene are shown in Fig. 4, which includes reactions that have been demonstrated in vivo or proposed from in vitro studies. An early study of the metabolic fate of inhaled isoprene used ^{14}C-labeled isoprene, which was inhaled by rats at one of four concentrations ranging from 8 to 8,200 ppm (Dahl et al. 1987). The study characterized disposition of the radiocarbon as 75% in urine and a very small percentage (0.0018%–0.031%) as a putative diepoxide metabolite. Metabolites in blood were characterized by vacuum line cryogenic distillation, which did not resolve the monoepoxide metabolites or the diepoxide from diol metabolites. Observation of metabolites in the respiratory tract at low blood concentrations were said to indicate that a substantial fraction of metabolism occurred in the respiratory tract.

Gervasi and coworkers noted that epoxide metabolites formed following incubation with rodent liver microsomes and that the diepoxide of isoprene (2-methyl-2,2′-bioxirane; see Fig. 4) was equivalent in its mutagenic potential to that of the diepoxide of butadiene, DEB (Del Monte et al. 1985; Gervasi et al. 1985; Gervasi and Longo 1990). Formation of isoprene epoxide metabolites has been attributed to CYP2E1, with minor activity by CYP2B6 (Bogaards et al. 1999).

Half-lives of the isoprene epoxides in aqueous media have been determined, and a distinction was noted in their stabilities. If the epoxide was formed on the methyl-substituted carbon (IP-1,2-O; Fig. 4) the half-life was 1.25 hr (Gervasi and Longo 1990), whereas half-lives of the other monoepoxide (IP-3,4-O) and the diepoxide were much longer, 73 hr and 46 hr, respectively. Instability of IP-1,2-O likely accounts for substantial metabolism

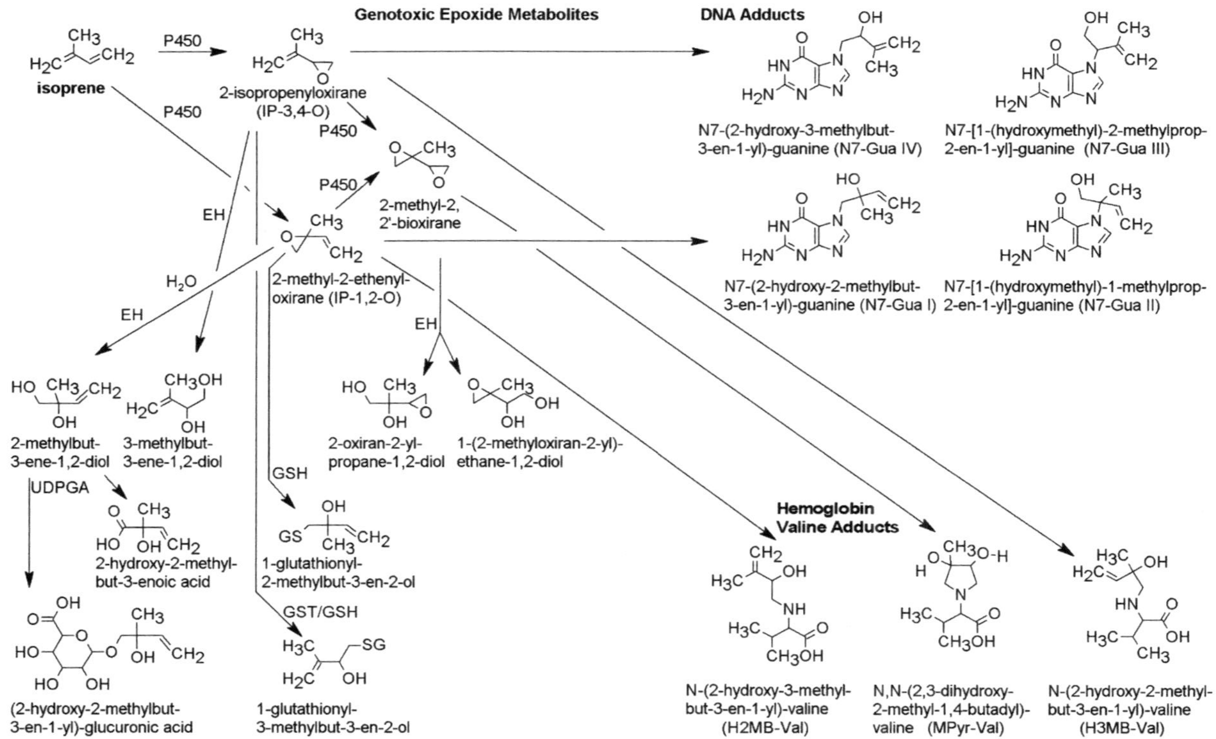

Fig. 4. Metabolic scheme for isoprene. *Note*: 1-hydroxy- and 2-hydroxy adduct regiosiomer pairs may exist from electrophile–nucleophile reactions that are not shown for simplicity.

to water-soluble metabolites, including a glucuronide of the hydrolysis product 2-methylbut-3-ene-1,2 diol and a corresponding oxidized metabolite, 2-hydroxy-2-methylbut-3-enoic acid. This instability resulted in postulation that IP-1,2-O did not lead to production of the diepoxide, but this conclusion appears to have been discounted (Watson et al. 2001; Wistuba et al. 1994). Detailed analyses of the stereoselectivity of metabolism in vitro of isoprene by rat (Chiappe et al. 2000) and mouse liver enzymes have been published (Wistuba et al. 1994), and comparative studies of stereoselectivity in liver microsomal metabolism have been accomplished for mice, rats, rabbits, dogs, monkeys, and humans (Small et al. 1997). Stereochemical effects as well as detailed in vitro kinetic parameters of mouse, rat, and human microsomal metabolism are available (Golding et al. 2003). A potentially important observation (Chiappe et al. 2000) that may have important implications for isoprene toxicity was that the diol epoxides were not hydrolyzed in rat liver microsomes. In these studies involving metabolism of isoprene epoxide stereoisomers by control and phenobarbital-induced microsomes, no hydrolysis by EH of diol epoxides was noted. In contrast, the isoprene diepoxide metabolite stereoisomers were metabolized by EH at a moderate rate in vitro. This study concluded that the 3,4-epoxy-2-methyl-1, 2-diol metabolite was most significant due to its rate of formation compared to other metabolites. However, assessment of the true significance of this metabolite and others in the genotoxicity of isoprene will require additional work, including assessment of DNA adducts and their mutagenicity.

D. Biomarkers

Adducts to N-terminal valine in hemoglobin have been detected following intraperitoneal injections of isoprene or isoprene epoxides to rats (Tareke et al. 1998). These studies used the modified Edman degradation method (Tornqvist et al. 1986, 2002) for adduct isolation, and gas chromatography/mass spectrometry (GC/MS) for detection and measurement. It is interesting to note that N-valine adducts were formed in vitro and in vivo with both monoepoxides of isoprene; this appears to call into question the assumption that these adducts are not generally mutagenic, based on bacterial assays (see following). These studies with the monoepoxides confirmed the rapid hydrolysis of IP-1,2-O as already noted, whereas IP-3,4-O was more persistent. Further studies using liquid chromatography/mass spectrometry (LC/MS) have studied N-valine adducts in model peptides with the diepoxide of isoprene (Fred et al. 2004a). These studies noted a cyclic valine adduct, MPyr-Val, which is homologous to the Pyr-Val adduct observed with butadiene diepoxide.

E. Genotoxicity

Exposure of mice to isoprene at 438, 1,750, or 7,000 ppm for 12 d induced significant increases in the frequency of sister chromatid exchange in bone marrow cells and in the levels of micronucleated polychromatic erythrocytes and of micronucleated normochromatic erythrocytes in peripheral blood at all concentrations (Tice et al. 1988). Significant lengthening of the average generation time in bone marrow and significant decrease in the percentage of circulating polychromatic erythrocytes were detected. Exposure to isoprene did not induce increase in frequency of chromosomal aberrations and did not alter the mitotic index in bone marrow. These results suggest that isoprene can induce tumors at multiple sites in B6C3F1 mice, as observed with butadiene.

Increases in micronucleated polychromatic erythrocytes, as a function of formed isoprene diepoxide metabolite, have been examined in conjunction with specific valine hemoglobin biomarkers following intraperitoneal dosage of IP-1,2-O (Fred et al. 2005). Increased frequencies of micronuclei formation in mice were observed with increased systemic doses of isoprene diepoxide, which was calculated from hemoglobin adduct levels and the previously determined rate constant of isoprene diepoxide reaction with hemoglobin. The calculated systemic diepoxide dosage was determined through use of the MPyr-Val hemoglobin adduct, which is formed by the diepoxide of isoprene.

In contrast to butadiene and chloroprene, both monoepoxides of isoprene (see Fig. 4; IP-1,2-O and IP-3,4-O) were reported to be nonmutagenic in the Ames *Salmonella typhimurium* assay (Gervasi et al. 1985), whereas the diepoxide of isoprene was noted in this bacterial system to be as mutagenic as the diepoxide of butadiene. It has been suggested (Watson et al. 2001) that the potential for isoprene diepoxide cross-linking between DNA bases is lower than that of butadiene diepoxide, in part due to a greater tendency to form an intramolecular cyclic adduct during reactions with the N^6-amino group of deoxyadenosine, rather than an interadenosine cross link. No reports concerning mutagenicity of the diol epoxide metabolites have been published. Reactions of isoprene monoepoxides with the N7-position of guanosine have been studied (Begemann et al. 2004), and the reaction products are included in Fig. 4. However, these adducts are not expected to be highly mutagenic, probably due to facile depurination and DNA repair (Begemann et al. 2004).

Harderian gland tumors produced during the 26-wk isoprene mouse inhalation study noted above were examined for mutations in K-*ras* and H-*ras* protooncogenes (Hong et al. 1997). These detailed studies identified mutations by single-strand conformational analysis and direct sequencing of polymerase chain reaction-amplified DNA. K-*ras* mutations were detected with higher frequency in these neoplasms than in spontaneous Harderian gland tumors or similar tumors obtained following butadiene inhalation studies in mice. All Harderian gland neoplasms induced by isoprene possessed activated *ras* genes, and of these 60% were K-*ras* and 40% were H-*ras* types. The predominant mutations induced by isoprene were A to T transversions, as CAA converted to CTA at K-*ras* codon 61, or C to A transversions, as CAA changed to AAA at H-*ras* codon 61. Hong et al. (1997) indicated that the high frequency of mutations and specificity of the mutation profile indicate that *ras* protooncogene activation probably contributed to carcinogenesis of the Harderian gland in these studies.

F. Human Epidemiology

Epidemiological studies of isoprene exposure in workers do not appear to have been conducted (Bird et al. 2001; IARC 1999c). Such studies would be complicated by the fact that isoprene is generated endogenously. Exposure assessment for isoprene, on which most reliable epidemiological studies are founded, would need to consider exogenous exposures concurrently with variable endogenous production.

Based on data from animal studies and homology with butadiene and chloroprene, evaluation of isoprene by IARC (1999c) indicated there is *sufficient evidence* in experimental animals for the carcinogenicity of isoprene, and that isoprene is *possibly* carcinogenic to humans (Group 2B). In the most recent Report on Carcinogens (NTP 2005c), the National Toxicology Program has classified isoprene as *reasonably anticipated to be a human carcinogen*, based on what has been deemed sufficient evidence of tumor formation at multiple organ sites in multiple species of experimental animals.

Summary

The diene monomers, 1,3-butadiene, chloroprene, and isoprene, respectively, differ only in substitution of a hydrogen, a chlorine, or a methyl group at the second of the four unsaturated carbon atoms in these linear molecules. Literature reviewed in the preceding sections indicates that these chemicals have important uses in synthesis of polymers, which offer significant benefits within modern society. Additionally, studies document that these monomers can increase the tumor formation rate in various organs of rats and mice during chronic cancer bioassays. The extent of tumor formation versus animal exposure to these monomers varies significantly across species, as well among strains within species. These studies approach, but do not resolve, important questions of human risk from inhalation exposure.

Each of these diene monomers can be activated to electrophilic epoxide metabolites through microsomal oxidation reactions in mammals. These epoxide metabolites are genotoxic through reactions with nucleic acids. Some of these reactions cause mutations and subsequent cancers, as noted in animal experiments. Significant differences exist among the compounds, particularly in the extent of formation of highly mutagenic diepoxide metabolites, when animals are exposed. These metabolites are detoxified through hydrolysis by epoxide hydrolase enzymes and through conjugation with glutathione with the aid of glutathione *S*-transferase. Different strains and species perform these reactions with varying efficacy. Mice produce these electrophilic epoxides more rapidly and appear to have less adequate detoxification mechanisms than rats or humans. The weight of evidence from many studies suggests that the balance of activation versus detoxification offers explanation of differing sensitivities of animals to these carcinogenic actions. Other aspects, including molecular biology of the many processes that lead through specific mutations to cancer, are yet to be understood.

Melnick and Sills (2001) compared the carcinogenic potentials of these three dienes, along with that of ethylene oxide, which also acts through an epoxide intermediate. From the number of tissue sites where experimental animal tumors were detected, butadiene offers greatest potential for carcinogenicity of these dienes. Chloroprene and then isoprene appear to follow in this order. Comparisons among these chemicals based on responses to external exposures are complicated by differences among studies and of species and tissue susceptibilities. Physiologically based pharmacokinetic models offer promise to overcome these impediments to interpretation. Mechanistic studies at the molecular level offer promise for understanding the relationships among electrophilic metabolites and vital genetic components.

Significant improvements in minimization of industrial worker exposures to carcinogenic chemicals have been accomplished after realization that vinyl chloride caused hepatic angiosarcoma in polymer production workers (Creech and Johnson 1974; Falk et al. 1974). Efforts continue to minimize disease, particularly cancer, from exposures to chemicals such as these dienes. Industry has responded to significant challenges that affect the health of workers through efforts that minimize plant exposures and by sponsorship of research, including animal and epidemiological studies. Governmental agencies provide oversight and have developed facilities that accomplish studies of continuing scientific excellence. These entities grapple with differences in perspective, objectives, and interpretation as synthesis of knowledge develops through mutual work. A major challenge remains, however, in assessment of significance of environmental human exposures to these dienes. Such exposure levels are orders of magnitude less than exposures studied in experimental or epidemiological settings, but exposures may persist much longer and may involve unknown but potentially significant sensitivities in the general population. New paradigms likely will be needed for toxicological evaluation of these human exposures, which are ongoing but as yet are not interpreted.

References

Abdel-Rahman, SZ, Ammenheuser, MM, Ward, JB Jr (2001) Human sensitivity to 1,3-butadiene: role of microsomal epoxide hydrolase polymorphisms. Carcinogenesis (Oxf) 22:415–423.

ACC Olefins Panel (2005) HPV Category Summary: Pyrolysis C3+ and Pyrolysis C4+ Category. Report 201-15092. American Chemistry Council, Arlington, VA.

Acquavella, JF, Leonard, RC (2001) A review of the epidemiology of 1,3-butadiene and chloroprene. Chem-Biol Interact 135–136:43–52.

Affek, HP, Yakir, D (2002) Protection by isoprene against singlet oxygen in leaves. Plant Physiol 129:269–277.

Albertini, RJ, Sram, RJ, Vacek, PM, Lynch, J, Nicklas, JA, van Sittert, NJ, Boogaard, PJ, Henderson, RF, Swenberg, JA, Tates, AD, Ward, JB Jr, Wright, M, Ammenheuser, MM, Binkova, B, Blackwell, W, de Zwart, FA, Krako, D, Krone, J, Megens, H, Musilova, P, Rajska, G, Ranasinghe, A, Rosenblatt, JI, Rossner, P, Rubes, J, Sullivan, L, Upton, P, Zwinderman, AH (2003) Biomarkers in Czech workers exposed to 1,3-butadiene: a transitional epidemiologic study. Res Rep Health Eff Inst 116. Health Effects Institute, Boston, MA, pp 1–141.

Ames, BN (1973) Carcinogens are mutagens: their detection and classification. Environ Health Perspect 6:115–118.

Ames, BN, Durston, WE, Yamasaki, E, Lee, FD (1973a) Carcinogens are mutagens: a simple test system combining liver homogenates for activation and bacteria for detection. Proc Natl Acad Sci USA 70:2281–2285.

Ames, BN, Lee, FD, Durston, WE (1973b) An improved bacterial test system for the detection and classification of mutagens and carcinogens. Proc Natl Acad Sci USA 70:782–786.

Bartsch, H, Malaveille, C, Montesano, R, Tomatis, L (1975) Tissue-mediated mutagenicity of vinylidene chloride and 2-chlorobutadiene in *Salmonella typhimurium*. Nature (Lond) 255:641–643.

Bartsch, H, Malaveille, C, Barbin, A, Bresil, H, Tomatis, L, Montesano, R (1976) Mutagenicity and metabolism of vinyl chloride and related compounds. Environ Health Perspect 17:193–198.

Bartsch, H, Malaveille, C, Barbin, A, Planche, G (1979) Mutagenic and alkylating metabolites of halo-ethylenes, chlorobutadienes and dichlorobutenes produced by rodent or human liver tissues. Evidence for oxirane formation by P450-linked microsomal mono-oxygenases. Arch Toxicol 41:249–277.

Bechtold, WE, Strunk, MR, Chang, IY, Ward, JB Jr, Henderson, RF (1994) Species differences in urinary butadiene metabolites: comparisons of metabolite ratios between mice, rats, and humans. Toxicol Appl Pharmacol 127:44–49.

Begemann, P, Christova-Georgieva, NI, Sangaiah, R, Koc, H, Zhang, D, Golding, BT, Gold, A, Swenberg, JA (2004) Synthesis, characterization, and identification of N7-guanine adducts of isoprene monoepoxides in vitro. Chem Res Toxicol 17:929–936.

Ben Porath, I, Weinberg, RA (2005) The signals and pathways activating cellular senescence. Int J Biochem Cell Biol 37:961–976.

Bennett, DA, Waters, MD (2000) Applying biomarker research. Environ Health Perspect 108:907–910.

Bird, MG, Rice, JM, Bond, JA (2001) Evaluation of 1,3-butadiene, isoprene and chloroprene health risks. Chem-Biol Interact 135–136:1–7.

Blair, IA, Oe, T, Kambouris, S, Chaudhary, AK (2000) 1,3-Butadiene: cancer, mutations, and adducts. Part IV: Molecular dosimetry of 1,3-butadiene. Res Rep Health Eff Inst 92. Health Effects Institute, Boston, MA, pp 151–190.

Bogaards, JJ, Venekamp, JC, Salmon, FG, van Bladeren, PJ (1999) Conjugation of isoprene monoepoxides with glutathione, catalyzed by alpha, mu, pi and theta-class glutathione S-transferases of rat and man. Chem-Biol Interact 117: 1–14.

Bolt, HM (1986) Metabolic activation of vinyl chloride, formation of nucleic acid adducts and relevance to carcinogenesis. IARC Sci Publ 70:261–268.

Bolt, HM, Schmiedel, G, Filser, JG, Rolzhauser, HP, Lieser, K, Wistuba, D, Schurig, V (1983) Biological activation of 1,3-butadiene to vinyl oxirane by rat liver microsomes and expiration of the reactive metabolite by exposed rats. J Cancer Res Clin Oncol 106:112–116.

Bond, JA, Medinsky, MA (2001) Insights into the toxicokinetics and toxicodynamics of 1,3-butadiene. Chem-Biol Interact 135–136:599–614.

Bond, JA, Himmelstein, MW, Seaton, M, Boogaard, P, Medinsky, MA (1996) Metabolism of butadiene by mice, rats, and humans: a comparison of physiologically based toxicokinetic model predictions and experimental data. Toxicology 113:48–54.

Boogaard, PJ, Bond, JA (1996) The role of hydrolysis in the detoxification of 1,2:3,4-diepoxybutane by human, rat, and mouse liver and lung in vitro. Toxicol Appl Pharmacol 141:617–627.

Boogaard, PJ, de Kloe, KP, Booth, ED, Watson, WP (2004) DNA adducts in rats and mice following exposure to [4-14C]-1,2-epoxy-3-butene and to [2,3-14C]-1,3-butadiene. Chem-Biol Interact 148:69–92.

Booth, ED, Kilgour, JD, Robinson, SA, Watson, WP (2004) Dose responses for DNA adduct formation in tissues of rats and mice exposed by inhalation to low concentrations of 1,3-[2,3-[(14)C]-butadiene. Chem-Biol Interact 147:195–211.

Boysen, G, Georgieva, NI, Upton, PB, Jayaraj, K, Li, Y, Walker, VE, Swenberg, JA (2004) Analysis of diepoxide-specific cyclic N-terminal globin adducts in mice and rats after inhalation exposure to 1,3-butadiene. Cancer Res 64:8517–8520.

California Air Resources Board (2006) 1,3-Butadiene at John Swett. http://www.arb.ca.gov/ch/aq_result/crockett/cr_buta.htm (3-10-2006).

Carmical, JR, Nechev, LV, Harris, CM, Harris, TM, Lloyd, RS (2000a) Mutagenic potential of adenine N(6) adducts of monoepoxide and diolepoxide derivatives of butadiene. Environ Mol Mutagen 35:48–56.

Carmical, JR, Zhang, M, Nechev, L, Harris, CM, Harris, TM, Lloyd, RS (2000b) Mutagenic potential of guanine N2 adducts of butadiene mono- and diolepoxide. Chem Res Toxicol 13:18–25.

CEN (2005) Facts & figures from the chemical industry. http://pubs.acs.org/cen/coverstory/83/8328factsandfigures.html (4-10-2006).

Chiappe, C, De Rubertis, A, Tinagli, V, Amato, G, Gervasi, PG (2000) Stereochemical course of the biotransformation of isoprene monoepoxides and of the corresponding diols with liver microsomes from control and induced rats. Chem Res Toxicol 13:831–838.

Clary, JJ, Feron, VJ, Reuzel, PG (1978) Toxicity of beta-chloroprene (2-chlorobutadiene-1,3): acute and subacute toxicity. Toxicol Appl Pharmacol 46:375–384.

Cochrane, JE, Skopek, TR (1994a) Mutagenicity of butadiene and its epoxide metabolites: I. Mutagenic potential of 1,2-epoxybutene, 1,2,3,4-diepoxybutane and 3,4-epoxy-1,2-butanediol in cultured human lymphoblasts. Carcinogenesis (Oxf) 15:713–717.

Cochrane, JE, Skopek, TR (1994b) Mutagenicity of butadiene and its epoxide metabolites: II. Mutational spectra of butadiene, 1,2-epoxybutene and diepoxy-

butane at the hprt locus in splenic T cells from exposed B6C3F1 mice. Carcinogenesis (Oxf) 15:719–723.
Coon, MJ (2005) Cytochrome P450: nature's most versatile biological catalyst. Annu Rev Pharmacol Toxicol 45:1–25.
Cottrell, L, Golding, BT, Munter, T, Watson, WP (2001) In vitro metabolism of chloroprene: species differences, epoxide stereochemistry and a de-chlorination pathway. Chem Res Toxicol 14:1552–1562.
Creech, JL Jr, Johnson, MN (1974) Angiosarcoma of liver in the manufacture of polyvinyl chloride. J Occup Med 16:150–151.
Csanady, GA, Guengerich, FP, Bond, JA (1992) Comparison of the biotransformation of 1,3-butadiene and its metabolite, butadiene monoepoxide, by hepatic and pulmonary tissues from humans, rats and mice. Carcinogenesis (Oxf) 13:1143–1153.
Csanady, GA, Kreuzer, PE, Baur, C, Filser, JG (1996) A physiological toxicokinetic model for 1,3-butadiene in rodents and man: blood concentrations of 1,3-butadiene, its metabolically formed epoxides, and of haemoglobin adducts – relevance of glutathione depletion. Toxicology 113:300–305.
Dahl, AR, Birnbaum, LS, Bond, JA, Gervasi, PG, Henderson, RF (1987) The fate of isoprene inhaled by rats: comparison to butadiene. Toxicol Appl Pharmacol 89:237–248.
Del Monte, M, Citti, L, Gervasi, PG (1985) Isoprene metabolism by liver microsomal mono-oxygenases. Xenobiotica 15:591–597.
Delzell, E, Sathiakumar, N, Macaluso, M, Hovinga, M, Larson, R, Barone, F, Beall, C, Julian, J, Muir, D (1995) A follow-up study of synthetic rubber workers. Prepared for the International Institute of Synthetic Rubber Workers, Houston, TX.
Delzell, E, Sathiakumar, N, Hovinga, M, Macaluso, M, Julian, J, Larson, R, Cole, P, Muir, DC (1996) A follow-up study of synthetic rubber workers. Toxicology 113:182–189.
Drevon, C, Kuroki, T (1979) Mutagenicity of vinyl chloride, vinylidene chloride and chloroprene in V79 Chinese hamster cells. Mutat Res 67:173–182.
Duescher, RJ, Elfarra, AA (1994) Human liver microsomes are efficient catalysts of 1,3-butadiene oxidation: evidence for major roles by cytochromes P450 2A6 and 2E1. Arch Biochem Biophys 311:342–349.
Elfarra, AA, Sharer, JE, Duescher, RJ (1995) Synthesis and characterization of *N*-acetyl-L-cysteine S-conjugates of butadiene monoxide and their detection and quantitation in urine of rats and mice given butadiene monoxide. Chem Res Toxicol 8:68–76.
Elfarra, AA, Krause, RJ, Selzer, RR (1996) Biochemistry of 1,3-butadiene metabolism and its relevance to 1,3-butadiene-induced carcinogenicity. Toxicology 113:23–30.
Elfarra, AA, Krause, RJ, Kemper, RA (2001a) Cellular and molecular basis for species, sex and tissue differences in 1,3-butadiene metabolism. Chem-Biol Interact 135–136:239–248.
Elfarra, AA, Moll, TS, Krause, RJ, Kemper, RA, Selzer, RR (2001b) Reactive metabolites of 1,3-butadiene: DNA and hemoglobin adduct formation and potential roles in carcinogenicity. Adv Exp Med Biol 500:93–103.
Evans, T, Hart, MJ, Cerione, RA (1991) The Ras superfamilies: regulatory proteins and posttranslational modifications. Curr Opin Cell Biol 3:185–191.

Falk, H, Creech, JL Jr, Heath, CW Jr, Johnson, MN, Key, MM (1974) Hepatic disease among workers at a vinyl chloride polymerization plant. JAMA 230:59–63.
Filser, JG, Johanson, G, Kessler, W, Kreuzer, PE, Stei, P, Baur, C, Csanady, GA (1993) A pharmacokinetic model to describe toxicokinetic interactions between 1,3-butadiene and styrene in rats: predictions for human exposure. IARC Sci Publ 127:65–78.
Fred, C, Cantillana, T, Henderson, AP, Golding, BT, Tornqvist, M (2004a) Adducts of N-terminal valines in hemoglobin with isoprene diepoxide, a metabolite of isoprene. Rapid Commun Mass Spectrom 18:2177–2184.
Fred, C, Kautiainen, A, Athanassiadis, I, Tornqvist, M (2004b) Hemoglobin adduct levels in rat and mouse treated with 1,2:3,4-diepoxybutane. Chem Res Toxicol 17:785–794.
Fred, C, Grawe, J, Tornqvist, M (2005) Hemoglobin adducts and micronuclei in rodents after treatment with isoprene monoxide or butadiene monoxide. Mutat Res 585:21–32.
Fustinoni, S, Soleo, L, Warholm, M, Begemann, P, Rannug, A, Neumann, HG, Swenberg, JA, Vimercati, L, Colombi, A (2002) Influence of metabolic genotypes on biomarkers of exposure to 1,3-butadiene in humans. Cancer Epidemiol Biomark Prev 11:1082–1090.
Gervasi, PG, Longo, V (1990) Metabolism and mutagenicity of isoprene. Environ Health Perspect 86:85–87.
Gervasi, PG, Citti, L, Del Monte, M, Longo, V, Benetti, D (1985) Mutagenicity and chemical reactivity of epoxidic intermediates of the isoprene metabolism and other structurally related compounds. Mutat Res 156:77–82.
Golding, BT, Cottrell, L, Mackay, D, Zhang, D, Watson, WP (2003) Stereochemical and kinetic comparisons of mono- and diepoxide formation in the in vitro metabolism of isoprene by liver microsomes from rats, mice, and humans. Chem Res Toxicol 16:933–944.
Goodrow, T, Reynolds, S, Maronpot, R, Anderson, M (1990) Activation of K-ras by codon 13 mutations in C57BL/6 X C3H F1 mouse tumors induced by exposure to 1,3-butadiene. Cancer Res 50:4818–4823.
Graff, JJ, Sathiakumar, N, Macaluso, M, Maldonado, G, Matthews, R, Delzell, E (2005) Chemical exposures in the synthetic rubber industry and lymphohematopoietic cancer mortality. J Occup Environ Med 47:916–932.
Guengerich, FP (1997) Comparisons of catalytic selectivity of cytochrome P450 subfamily enzymes from different species. Chem-Biol Interact 106:161–182.
Guengerich, FP (2003) Cytochrome P450 oxidations in the generation of reactive electrophiles: epoxidation and related reactions. Arch Biochem Biophys 409:59–71.
Gwinner, LM, Laib, RJ, Filser, JG, Bolt, HM (1983) Evidence of chloroethylene oxide being the reactive metabolite of vinyl chloride towards DNA: comparative studies with 2,2′-dichlorodiethylether. Carcinogenesis (Oxf) 4:1483–1486.
Hayes, JD, Flanagan, JU, Jowsey, IR (2005) Glutathione transferases. Annu Rev Pharmacol Toxicol 45:51–88.
Hayes, RB, Zhang, L, Yin, S, Swenberg, JA, Xi, L, Wiencke, J, Bechtold, WE, Yao, M, Rothman, N, Haas, R, O'Neill, JP, Zhang, D, Wiemels, J, Dosemeci, M, Li, G, Smith, MT (2000) Genotoxic markers among butadiene polymer workers in China. Carcinogenesis (OXF) 21:55–62.

Henderson, RF (2001) Species differences in the metabolism of olefins: implications for risk assessment. Chem-Biol Interact 135–136:53–64.

Henderson, RF, Barr, EB, Belinsky, SA, Benson, JM, Hahn, FF, Menache, MG (2000) 1,3-Butadiene: cancer, mutations, and adducts. Part I: Carcinogenicity of 1,2,3,4-diepoxybutane. Res Rep Health Eff Inst 92. Health Effects Institute, Boston, MA, pp 11–43.

Himmelstein, MW, Turner, MJ, Asgharian, B, Bond, JA (1994) Comparison of blood concentrations of 1,3-butadiene and butadiene epoxides in mice and rats exposed to 1,3-butadiene by inhalation. Carcinogenesis (Oxf) 15:1479–1486.

Himmelstein, MW, Asgharian, B, Bond, JA (1995) High concentrations of butadiene epoxides in livers and lungs of mice compared to rats exposed to 1,3-butadiene. Toxicol Appl Pharmacol 132:281–288.

Himmelstein, MW, Turner, MJ, Asgharian, B, Bond, JA (1996) Metabolism of 1,3-butadiene: inhalation pharmacokinetics and tissue dosimetry of butadiene epoxides in rats and mice. Toxicology 113:306–309.

Himmelstein, MW, Acquavella, JF, Recio, L, Medinsky, MA, Bond, JA (1997) Toxicology and epidemiology of 1,3-butadiene. Crit Rev Toxicol 27:1–108.

Himmelstein, MW, Carpenter, SC, Hinderliter, PM, Snow, TA, Valentine, R (2001a) The metabolism of beta-chloroprene: preliminary in-vitro studies using liver microsomes. Chem-Biol Interact 135–136:267–284.

Himmelstein, MW, Gladnick, NL, Donner, EM, Snyder, RD, Valentine, R (2001b) In vitro genotoxicity testing of (1-chloroethenyl)oxirane, a metabolite of beta-chloroprene. Chem-Biol Interact 135–136:703–713.

Himmelstein, MW, Carpenter, SC, Evans, MV, Hinderliter, PM, Kenyon, EM (2004a) Kinetic modeling of beta-chloroprene metabolism: II. the application of physiologically based modeling for cancer dose response analysis. Toxicol Sci 79(1):28–37.

Himmelstein, MW, Carpenter, SC, Hinderliter, PM (2004b) Kinetic Modeling of beta-chloroprene metabolism: I. In vitro rates in liver and lung tissue fractions from mice, rats, hamsters, and humans. Toxicol Sci 79(1):18–27.

Hirabayashi, Y (2005) p53-dependent gene profiling for reactive oxygen species after benzene inhalation: special reference to genes associated with cell cycle regulation. Chem-Biol Interact 153–154:165–170.

Hong, HL, Devereux, TR, Melnick, RL, Eldridge, SR, Greenwell, A, Haseman, J, Boorman, GA, Sills, RC (1997) Both K-ras and H-ras protooncogene mutations are associated with Harderian gland tumorigenesis in B6C3F1 mice exposed to isoprene for 26 weeks. Carcinogenesis (Oxf) 18:783–789.

Hong, HL, Devereux, TR, Melnick, RL, Moomaw, CR, Boorman, GA, Sills, RC (2000) Mutations of ras protooncogenes and p53 tumor suppressor gene in cardiac hemangiosarcomas from B6C3F1 mice exposed to 1,3-butadiene for 2 years. Toxicol Pathol 28:529–534.

Howard, P, Boethling, R, Jarvis, W, Meylan, W, Michalenko, E (1991) Handbook of Environmental Degradation Rates. Lewis, Chelsea, MI.

Huff, JE, Melnick, RL, Solleveld, HA, Haseman, JK, Powers, M, Miller, RA (1985) Multiple organ carcinogenicity of 1,3-butadiene in B6C3F1 mice after 60 weeks of inhalation exposure. Science 227:548–549.

Hughes, K, Meek, ME, Walker, M, Beauchamp, R (2001) 1,3-Butadiene: Human Health Aspects. Concise International Chemical Assessment Doc 30. Health Canada, Ottawa, Ontario, pp 1–73.

Hughes, K, Meek, ME, Walker, M, Beauchamp, R (2003) 1,3-Butadiene: exposure estimation, hazard characterization, and exposure-response analysis. J Toxicol Environ Health B Crit Rev 6:55–83.
Hurst, HE, Ali, MY (2006) Analyses of (1-chloroethenyl)oxirane headspace and hemoglobin N-valine adducts in erythrocytes indicate selective detoxification of (1-chloroethenyl)oxirane enantiomers. Chem-Biol Interact (in press).
IARC (1979) Chloroprene and polychloroprene. IARC Monographs on the Evaluation of Carcinogenic Risks to Humans, vol 19. World Health Organization, International Agency for Research on Cancer, Lyon, France, pp 131–156.
IARC (1999a) 1,3-Butadiene. IARC Monographs on the Evaluation of Carcinogenic Risks to Humans, vol 71, pt 1. World Health Organization, International Agency for Research on Cancer, Lyon, France, pp 109–225.
IARC (1999b) Chloroprene. IARC Monographs on the Evaluation of Carcinogenic Risks to Humans, vol 71, pt 1. World Health Organization, International Agency for Research on Cancer, Lyon, France, pp 227–250.
IARC (1999c) Isoprene. IARC Monographs on the Evaluation of Carcinogenic Risks to Humans, vol 71, pt 3. World Health Organization, International Agency for Research on Cancer, Lyon, France, pp 1015–1025.
Irons, RD, Smith, CN, Stillman, WS, Shah, RS, Steinhagen, WH, Leiderman, LJ (1986a) Macrocytic-megaloblastic anemia in male B6C3F1 mice following chronic exposure to 1,3-butadiene. Toxicol Appl Pharmacol 83:95–100.
Irons, RD, Smith, CN, Stillman, WS, Shah, RS, Steinhagen, WH, Leiderman, LJ (1986b) Macrocytic-megaloblastic anemia in male NIH Swiss mice following repeated exposure to 1,3-butadiene. Toxicol Appl Pharmacol 85:450–455.
Jackson, MA, Stack, HF, Rice, JM, Waters, MD (2000a) A review of the genetic and related effects of 1,3-butadiene in rodents and humans. Mutat Res 463:181–213.
Jackson, TE, Lilly, PD, Recio, L, Schlosser, PM, Medinsky, MA (2000b) Inhibition of cytochrome P450 2E1 decreases, but does not eliminate, genotoxicity mediated by 1,3-butadiene. Toxicol Sci 55:266–273.
Jacobson-Kram, D, Rosenthal, SL (1995) Molecular and genetic toxicology of 1,3-butadiene. Mutat Res 339:121–130.
Johanson, G, Filser, JG (1993) A physiologically based pharmacokinetic model for butadiene and its metabolite butadiene monoxide in rat and mouse and its significance for risk extrapolation. Arch Toxicol 67:151–163.
Johanson, G, Filser, JG (1996) PBPK model for butadiene metabolism to epoxides: quantitative species differences in metabolism. Toxicology 113:40–47.
Karl, T, Prazeller, P, Mayr, D, Jordan, A, Rieder, J, Fall, R, Lindinger, W (2001) Human breath isoprene and its relation to blood cholesterol levels: new measurements and modeling. J Appl Physiol 91:762–770.
Kemper, RA, Elfarra, AA, Myers, SR (1998) Metabolism of 3-butene-1,2-diol in B6C3F1 mice. Evidence for involvement of alcohol dehydrogenase and cytochrome p450. Drug Metab Dispos 26:914–920.
Kemper, RA, Krause, RJ, Elfarra, AA (2001) Metabolism of butadiene monoxide by freshly isolated hepatocytes from mice and rats: different partitioning between oxidative, hydrolytic, and conjugation pathways. Drug Metab Dispos 29:830–836.
Kligerman, AD, DeMarini, DM, Doerr, CL, Hanley, NM, Milholland, VS, Tennant, AH (1999) Comparison of cytogenetic effects of 3,4-epoxy-1-butene and

1,2:3,4-diepoxybutane in mouse, rat and human lymphocytes following in vitro G_0 exposures. Mutat Res 439:13–23.
Koc, H, Tretyakova, NY, Walker, VE, Henderson, RF, Swenberg, JA (1999) Molecular dosimetry of N-7 guanine adduct formation in mice and rats exposed to 1,3-butadiene. Chem Res Toxicol 12:566–574.
Kohn, MC (1997) The importance of anatomical realism for validation of physiological models of disposition of inhaled toxicants. Toxicol Appl Pharmacol 147:448–458.
Kohn, MC, Melnick, RL (1993) Species differences in the production and clearance of 1,3-butadiene metabolites: a mechanistic model indicates predominantly physiological, not biochemical, control. Carcinogenesis (Oxf) 14:619–628.
Kohn, MC, Melnick, RL (1996) Effects of the structure of a toxicokinetic model of butadiene inhalation exposure on computed production of carcinogenic intermediates. Toxicology 113:31–39.
Kohn, MC, Melnick, RL (2000) The privileged access model of 1,3-butadiene disposition. Environ Health Perspect 108(suppl 5):911–917.
Koivisto, P, Kilpelainen, I, Rasanen, I, Adler, ID, Pacchierotti, F, Peltonen, K (1999) Butadiene diolepoxide- and diepoxybutane-derived DNA adducts at N7-guanine: a high occurrence of diolepoxide-derived adducts in mouse lung after 1,3-butadiene exposure. Carcinogenesis (Oxf) 20:1253–1259.
Krause, RJ, Elfarra, AA (1997) Oxidation of butadiene monoxide to meso- and (+/−)-diepoxybutane by cDNA-expressed human cytochrome P450s and by mouse, rat, and human liver microsomes: evidence for preferential hydration of meso-diepoxybutane in rat and human liver microsomes. Arch Biochem Biophys 337:176–184.
Krewski, D, Withey, JR, Ku, LF, Andersen, ME (1994) Applications of physiologic pharmacokinetic modeling in carcinogenic risk assessment. Environ Health Perspect 102(suppl 11):37–50.
Krewski, D, Henderson, RF, Bakshi, K (1999) Current trends in toxicological risk assessment: perspectives from the committee on toxicology. Inhal Toxicol 11:459–476.
Kumar, R, Sukumar, S, Barbacid, M (1990) Activation of ras oncogenes preceding the onset of neoplasia. Science 248:1101–1104.
Kuzma, J, Nemecek-Marshall, M, Pollock, WH, Fall, R (1995) Bacteria produce the volatile hydrocarbon isoprene. Curr Microbiol 30:97–103.
Kuzuyama, T, Seto, H (2003) Diversity of the biosynthesis of the isoprene units. Nat Prod Rep 20:171–183.
Landi, S (2000) Mammalian class theta GST and differential susceptibility to carcinogens: a review. Mutat Res 463:247–283.
Lawley, PD, Brookes, P (1967) Interstrand cross-linking of DNA by difunctional alkylating agents. J Mol Biol 25:143–160.
Leavens, TL, Bond, JA (1996) Pharmacokinetic model describing the disposition of butadiene and styrene in mice. Toxicology 113:310–313.
Leiderman, LJ, Stillman, WS, Shah, RS, Steinhagen, WH, Irons, RD (1986) Altered hematopoietic stem cell development in male B6C3F1 mice following exposure to 1,3-butadiene. Exp Mol Pathol 44:50–56.
Leonard, RC, Lineker, GA, Kreckmann, KH, Marsh, GM, Buchanich, JM, Youk, AO (2006) Comparison of standardized mortality ratios (SMRs) obtained from use of a reference population based on a company-wide registry

cohort to SMRs calculated against local and national rates. Chem-Biol Interact (in press).
Li, SQ, Dong, QN, Liu, YQ, Liu, YG (1989) Epidemiologic study of cancer mortality among chloroprene workers. Biomed Environ Sci 2:141–149.
Lynch, J (2001a) BD monomer and elastomer production processes. Chem-Biol Interact 135–136:147–153.
Lynch, J (2001b) Occupational exposure to butadiene, isoprene and chloroprene. Chem-Biol Interact 135–136:207–214.
Lynch, M (2001c) Manufacture and use of chloroprene monomer. Chem-Biol Interact 135–136:155–167.
Macaluso, M, Larson, R, Delzell, E, Sathiakumar, N, Hovinga, M, Julian, J, Muir, D, Cole, P (1996) Leukemia and cumulative exposure to butadiene, styrene and benzene among workers in the synthetic rubber industry. Toxicology 113:190–202.
Marsh, GM, Youk, AO, Buchanich, JM, Cunningham, M, Esmen, NA, Hall, TA, Phillips, ML (2006a) Mortality patterns among industrial workers exposed to chloroprene and other substances: I. General mortality patterns. Chem-Biol Interact (in press).
Marsh, GM, Youk, AO, Buchanich, JM, Cunningham, M, Esmen, NA, Hall, TA, Phillips, ML (2006b) Mortality patterns among industrial workers exposed to chloroprene and other substances: II. Mortality in relation to exposure. Chem-Biol Interact (in press).
Maryland Department of the Environment (2006) MDE Air Quality Data Report. http://www.mde.state.md.us/Programs/AirPrograms/airData/dataReport.asp (4-10-2006).
Medinsky, MA, Leavens, TL, Csanady, GA, Gargas, ML, Bond, JA (1994) In vivo metabolism of butadiene by mice and rats: a comparison of physiological model predictions and experimental data. Carcinogenesis (Oxf) 15:1329–1340.
Melnick, RL (2002) Carcinogenicity and mechanistic insights on the behavior of epoxides and epoxide-forming chemicals. Ann NY Acad Sci 982:177–189.
Melnick, RL, Huff, J (1992) 1,3-Butadiene: toxicity and carcinogenicity in laboratory animals and in humans. Rev Environ Contam Toxicol 124:111–144.
Melnick, RL, Kohn, MC (2000) Dose-response analyses of experimental cancer data. Drug Metab Rev 32:193–209.
Melnick, RL, Sills, RC (2001) Comparative carcinogenicity of 1,3-butadiene, isoprene, and chloroprene in rats and mice. Chem-Biol Interact 135–136:27–42.
Melnick, RL, Huff, J, Chou, BJ, Miller, RA (1990a) Carcinogenicity of 1,3-butadiene in C57BL/6 × C3H F1 mice at low exposure concentrations. Cancer Res 50: 6592–6599.
Melnick, RL, Huff, JE, Roycroft, JH, Chou, BJ, Miller, RA (1990b) Inhalation toxicology and carcinogenicity of 1,3-butadiene in B6C3F1 mice following 65 weeks of exposure. Environ Health Perspect 86:27–36.
Melnick, RL, Sills, RC, Roycroft, JH, Chou, BJ, Ragan, HA, Miller, RA (1994) Isoprene, an endogenous hydrocarbon and industrial chemical, induces multiple organ neoplasia in rodents after 26 weeks of inhalation exposure. Cancer Res 54:5333–5339.
Melnick, RL, Elwell, MR, Roycroft, JH, Chou, BJ, Ragan, HA, Miller, RA (1996) Toxicity of inhaled chloroprene (2-chloro-1,3-butadiene) in F344 rats and B6C3F(1) mice. Toxicology 108:79–91.

Melnick, RL, Sills, RC, Portier, CJ, Roycroft, JH, Chou, BJ, Grumbein, SL, Miller, RA (1999a) Multiple organ carcinogenicity of inhaled chloroprene (2-chloro-1,3-butadiene) in F344/N rats and B6C3F1 mice and comparison of dose-response with 1,3-butadiene in mice. Carcinogenesis (Oxf) 20:867–878.

Melnick, RL, Sills, RC, Portier, CJ, Roycroft, JH, Chou, BJ, Grumbein, SL, Miller, RA (1999b) Multiple organ carcinogenicity of inhaled chloroprene (2-chloro-1,3-butadiene) in F344/N rats and B6C3F1 mice and comparison of dose-response with 1,3-butadiene in mice. Carcinogenesis (Oxf) 20(5):867–878.

Merck Index (1996) Isoprene. In: Budavari S (ed) The Merck Index. Merck, Whitehouse Station, NJ, p 887.

Merritt, WK, Scholdberg, TA, Nechev, LV, Harris, TM, Harris, CM, Lloyd, RS, Stone, MP (2004) Stereospecific structural perturbations arising from adenine N(6) butadiene triol adducts in duplex DNA. Chem Res Toxicol 17:1007–1019.

Merritt, WK, Nechev, LV, Scholdberg, TA, Dean, SM, Kiehna, SE, Chang, JC, Harris, TM, Harris, CM, Lloyd, RS, Stone, MP (2005) Structure of the 1,4-bis(2′-deoxyadenosin-N6-yl)-2R,3R-butanediol cross-link arising from alkylation of the human N-ras codon 61 by butadiene diepoxide. Biochemistry-US 44:10081–10092.

Moll, TS, Elfarra, AA (1999) Characterization of the reactivity, regioselectivity, and stereoselectivity of the reactions of butadiene monoxide with valinamide and the N-terminal valine of mouse and rat hemoglobin. Chem Res Toxicol 12:679–689.

Moll, TS, Harms, AC, Elfarra, AA (2000) A comprehensive structural analysis of hemoglobin adducts formed after in vitro exposure of erythrocytes to butadiene monoxide. Chem Res Toxicol 13:1103–1113.

Morisseau, C, Hammock, BD (2005) Epoxide hydrolases: mechanisms, inhibitor designs, and biological roles. Annu Rev Pharmacol Toxicol 45:311–333.

Morrissey, RE, Schwetz, BA, Hackett, PL, Sikov, MR, Hardin, BD, McClanahan, BJ, Decker, JR, Mast, TJ (1990) Overview of reproductive and developmental toxicity studies of 1,3-butadiene in rodents. Environ Health Perspect 86:79–84.

Morrow, NL (2001) Significance of 1,3-butadiene to the US air toxics regulatory effort. Chem-Biol Interact 135–136:137–143.

Munter, T, Cottrell, L, Hill, S, Kronberg, L, Watson, WP, Golding, BT (2002) Identification of adducts derived from reactions of (1-chloroethenyl)oxirane with nucleosides and calf thymus DNA. Chem Res Toxicol 15:1549–1560.

Munter, T, Cottrell, L, Golding, BT, Watson, WP (2003) Detoxication pathways involving glutathione and epoxide hydrolase in the in vitro metabolism of chloroprene. Chem Res Toxicol 16:1287–1297.

NTP (1998) Toxicology and carcinogenesis studies of chloroprene (CAS No. 126-99-8) in F344/N Rats and B6C3F1 mice (inhalation studies). Tech Rep Ser 467:1–372.

NTP (1999) NTP toxicology and carcinogenesis studies of isoprene (CAS No. 78-79-5) in F344/N rats (inhalation studies). Natl Toxicol Program Tech Rep Ser 486:1–176.

NTP (2005a) 1,3-Butadiene CAS No. 106-99-0. U.S. Department of Health and Human Services, Public Health Service, National Toxicology Program, Research Triangle Park, NC.

NTP (2005b) Chloroprene CAS No. 126-99-8. U.S. Department of Health and Human Services, Public Health Service, National Toxicology Program, Research Triangle Park, NC.

NTP (2005c) Isoprene CAS No. 78-79-5. U.S. Department of Health and Human Services, Public Health Service, National Toxicology Program, Research Triangle Park, NC.
Nystrom, AE (1948) Health hazards in the chloroprene rubber industry and their prevention. A clinical and experimental study, with special reference to chloroprene and its oxidation and polymerization products. Acta Med Scand 132:1–125.
Oe, T, Kambouris, SJ, Walker, VE, Meng, Q, Recio, L, Wherli, S, Chaudhary, AK, Blair, IA (1999) Persistence of N7-(2,3,4-trihydroxybutyl)guanine adducts in the livers of mice and rats exposed to 1,3-butadiene. Chem Res Toxicol 12:247–257.
OSHA (1996) Occupational exposure to 1,3-butadiene. Fed Reg 61:56746–56856.
OSHA (2005a) Occupational Exposure to 1,3-Butadiene. http://www.osha.gov/pls/oshaweb/owadisp.show_document?p_table=FEDERAL_REGISTER&p_id=13585 (10-19-2005a).
OSHA (2005b) Safety and Health Topics: Vinyl Chloride. http://www.osha.gov/dts/chemicalsampling/data/CH_275395.html (10-19-2005b).
Osterman-Golkar, S, Bond, JA (1996) Biomonitoring of 1,3-butadiene and related compounds. Environ Health Perspect 104(Suppl 5):907–915.
Osterman-Golkar, S, Kautiainen, A, Bergmark, E, Hakansson, K, Maki-Paakkanen, J (1991) Hemoglobin adducts and urinary mercapturic acids in rats as biological indicators of butadiene exposure. Chem-Biol Interact 80:291–302.
Osterman-Golkar, S, Peltonen, K, Anttinen-Klemetti, T, Landin, HH, Zorcec, V, Sorsa, M (1996) Haemoglobin adducts as biomarkers of occupational exposure to 1,3-butadiene. Mutagenesis 11:145–149.
Osterman-Golkar, SM, Moss, O, James, A, Bryant, MS, Turner, M, Bond, JA (1998) Epoxybutene-hemoglobin adducts in rats and mice: dose response for formation and persistence during and following long-term low-level exposure to butadiene. Toxicol Appl Pharmacol 150:166–173.
Owen, PE, Glaister, JR, Gaunt, IF, Pullinger, DH (1987) Inhalation toxicity studies with 1,3-butadiene. 3. Two year toxicity/carcinogenicity study in rats. Am Ind Hyg Assoc J 48:407–413.
Pell, S (1978) Mortality of workers exposed to chloroprene. J Occup Med 20:21–29.
Penn, A, Snyder, CA (1996) 1,3-Butadiene, a vapor phase component of environmental tobacco smoke, accelerates arteriosclerotic plaque development. Circulation 93:552–557.
Perez, HL, Lahdetie, J, Landin, H, Kilpelainen, I, Koivisto, P, Peltonen, K, Osterman-Golkar, S (1997) Haemoglobin adducts of epoxybutanediol from exposure to 1,3-butadiene or butadiene epoxides. Chem-Biol Interact 105:181–198.
Placke, ME, Griffis, L, Bird, M, Bus, J, Persing, RL, Cox, LA Jr (1996) Chronic inhalation oncogenicity study of isoprene in B6C3F1 mice. Toxicology 113:253–262.
Plugge, H, Jaeger, RJ (1979) Acute inhalation toxicity of 2-chloro-1,3-butadiene (chloroprene): effects on liver and lung. Toxicol Appl Pharmacol 50:565–572.
Powley, MW, Jayaraj, K, Gold, A, Ball, LM, Swenberg, JA (2003) 1,N2-propanodeoxyguanosine adducts of the 1,3-butadiene metabolite, hydroxymethylvinyl ketone. Chem Res Toxicol 16:1448–1454.
Powley, MW, Li, Y, Upton, PB, Walker, VE, Swenberg, JA (2005) Quantification of DNA and hemoglobin adducts of 3,4-epoxy-1,2-butanediol in rodents exposed to 3-butene-1,2-diol. Carcinogenesis (Oxf) 26:1573–1580.
Recio, L, Saranko, CJ, Steen, AM (2000) 1,3-Butadiene: cancer, mutations, and adducts. Part II: Roles of two metabolites of 1,3-butadiene in mediating its in

vivo genotoxicity. Res Rep Health Eff Inst 92. Health Effects Institute, Boston, MA, pp 49–87.
Reya, T, Clevers, H (2005) Wnt signaling in stem cells and cancer. Nature (Lond) 434:843–850.
Richardson, KA, Megens, HJ, Webb, JD, van Sittert, NJ (1996) Biological monitoring of butadiene exposure by measurement of haemoglobin adducts. Toxicology 113:112–118.
Richardson, KA, Peters, MM, Wong, BA, Megens, RH, van Elburg, PA, Booth, ED, Boogaard, PJ, Bond, JA, Medinsky, MA, Watson, WP, van Sittert, NJ (1999) Quantitative and qualitative differences in the metabolism of ^{14}C-1,3-butadiene in rats and mice: relevance to cancer susceptibility. Toxicol Sci 49:186–201.
Rodriguez, DA, Kowalczyk, A, Ward, JB Jr, Harris, CM, Harris, TM, Lloyd, RS (2001) Point mutations induced by 1,2-epoxy-3-butene N1 deoxyinosine adducts. Environ Mol Mutagen 38:292–296.
Rydberg, P, Magnusson, AL, Zorcec, V, Granath, F, Tornqvist, M (1996) Adducts to N-terminal valines in hemoglobin from butadiene metabolites. Chem-Biol Interact 101:193–205.
Sabourin, PJ, Burka, LT, Bechtold, WE, Dahl, AR, Hoover, MD, Chang, IY, Henderson, RF (1992) Species differences in urinary butadiene metabolites; identification of 1,2-dihydroxy-4-(*N*-acetylcysteinyl)butane, a novel metabolite of butadiene. Carcinogenesis (Oxf) 13:1633–1638.
Sanotskii, IV (1976) Aspects of the toxicology of chloroprene: immediate and long-term effects. Environ Health Perspect 17:85–93.
Sapkota, A, Halden, RU, Dominici, F, Groopman, JD, Buckley, TJ (2006) Urinary biomarkers of 1,3-butadiene in environmental settings using liquid chromatography isotope dilution tandem mass spectrometry. Chem-Biol Interact 160:70–79.
Sathiakumar, N, Delzell, E, Hovinga, M, Macaluso, M, Julian, JA, Larson, R, Cole, P, Muir, DC (1998) Mortality from cancer and other causes of death among synthetic rubber workers. Occup Environ Med 55:230–235.
Satoh, T, Kaziro, Y (1992) Ras in signal transduction. Semin Cancer Biol 3:169–177.
Scholdberg, TA, Nechev, LV, Merritt, WK, Harris, TM, Harris, CM, Lloyd, RS, Stone, MP (2004) Structure of a site-specific major groove (2*S*,3*S*)-*N*6-(2,3,4-trihydroxybutyl)-2′-deoxyadenosyl DNA adduct of butadiene diol epoxide. Chem Res Toxicol 17:717–730.
Scholdberg, TA, Nechev, LV, Merritt, WK, Harris, TM, Harris, CM, Lloyd, RS, Stone, MP (2005) Mispairing of a site specific major groove (2*S*,3*S*)-*N*6-(2,3,4-trihydroxybutyl)-2′-deoxyadenosyl DNA adduct of butadiene diol epoxide with deoxyguanosine: formation of a dA(anti):dG(anti) pairing interaction. Chem Res Toxicol 18:145–153.
Sharkey, TD (1996) Isoprene synthesis by plants and animals. Endeavour 20:74–78.
Shugaev, BB (1969) Concentrations of hydrocarbons in tissues as a measure of toxicity. Arch Environ Health 18:878–882.
Shuker, DE (2002) The enemy at the gates? DNA adducts as biomarkers of exposure to exogenous and endogenous genotoxic agents. Toxicol Lett 134:51–56.
Sills, RC, Boorman, GA, Neal, JE, Hong, HL, Devereux, TR (1999a) Mutations in ras genes in experimental tumours of rodents. IARC Sci Publ 146:55–86.
Sills, RC, Hong, HL, Melnick, RL, Boorman, GA, Devereux, TR (1999b) High frequency of codon 61 K-ras A $\rightarrow$ T transversions in lung and Harderian gland neoplasms of B6C3F1 mice exposed to chloroprene (2-chloro-1,3-butadiene) for 2

years, and comparisons with the structurally related chemicals isoprene and 1,3-butadiene. Carcinogenesis (Oxf) 20:657–662.
Silver, GM, Fall, R (1995) Characterization of aspen isoprene synthase, an enzyme responsible for leaf isoprene emission to the atmosphere. J Biol Chem 270: 13010–13016.
Small, RD, Golding, BT, Watson, WP (1997) Species differences in the stereochemistry of the metabolism of isoprene in vitro. Xenobiotica 27:1155–1164.
Sukumar, S (1990) An experimental analysis of cancer: role of ras oncogenes in multistep carcinogenesis. Cancer Cells 2:199–204.
Sun, JD, Dahl, AR, Bond, JA, Birnbaum, LS, Henderson, RF (1989) Characterization of hemoglobin adduct formation in mice and rats after administration of [^{14}C]butadiene or [^{14}C]isoprene. Toxicol Appl Pharmacol 100:86–95.
Sweeney, LM, Himmelstein, MW, Schlosser, PM, Medinsky, MA (1996) Physiologically based pharmacokinetic modeling of blood and tissue epoxide measurements for butadiene. Toxicology 113:318–321.
Sweeney, LM, Schlosser, PM, Medinsky, MA, Bond, JA (1997) Physiologically based pharmacokinetic modeling of 1,3-butadiene, 1,2-epoxy-3-butene, and 1,2:3,4-diepoxybutane toxicokinetics in mice and rats. Carcinogenesis (Oxf) 18:611–625.
Sweeney, LM, Himmelstein, MW, Gargas, ML (2001) Development of a preliminary physiologically based toxicokinetic (PBTK) model for 1,3-butadiene risk assessment. Chem-Biol Interact 135–136:303–322.
Swenberg, JA, Christova-Gueorguieva, NI, Upton, PB, Ranasinghe, A, Scheller, N, Wu, KY, Yen, TY, Hayes, R (2000) 1,3-Butadiene: cancer, mutations, and adducts. Part V: Hemoglobin adducts as biomarkers of 1,3-butadiene exposure and metabolism. Res Rep Health Eff Inst 92. Health Effects Institute, Boston, MA, pp 191–210.
Swenberg, JA, Koc, H, Upton, PB, Georguieva, N, Ranasinghe, A, Walker, VE, Henderson, R (2001) Using DNA and hemoglobin adducts to improve the risk assessment of butadiene. Chem-Biol Interact 135–136:387–403.
Taalman, RD (1996) Isoprene: background and issues. Toxicology 113:242–246.
Tareke, E, Golding, BT, Small, RD, Tornqvist, M (1998) Haemoglobin adducts from isoprene and isoprene monoepoxides. Xenobiotica 28:663–672.
Thornton-Manning, JR, Dahl, AR, Bechtold, WE, Griffith, WC Jr, Henderson, RF (1995) Disposition of butadiene monoepoxide and butadiene diepoxide in various tissues of rats and mice following a low-level inhalation exposure to 1,3-butadiene. Carcinogenesis (Oxf) 16:1723–1731.
Thornton-Manning, JR, Dahl, AR, Bechtold, WE, Henderson, RF (1996) Gender and species differences in the metabolism of 1,3-butadiene to butadiene monoepoxide and butadiene diepoxide in rodents following low-level inhalation exposures. Toxicology 113:322–325.
Tice, RR, Boucher, R, Luke, CA, Shelby, MD (1987) Comparative cytogenetic analysis of bone marrow damage induced in male B6C3F1 mice by multiple exposures to gaseous 1,3-butadiene. Environ Mutagen 9:235–250.
Tice, RR, Boucher, R, Luke, CA, Paquette, DE, Melnick, RL, Shelby, MD (1988) Chloroprene and isoprene: cytogenetic studies in mice. Mutagenesis 3:141–146.
Tornqvist, M, Landin, HH (1995) Hemoglobin adducts for in vivo dose monitoring and cancer risk estimation. J Occup Environ Med 37:1077–1085.
Tornqvist, M, Mowrer, J, Jensen, S, Ehrenberg, L (1986) Monitoring of environmental cancer initiators through hemoglobin adducts by a modified Edman degradation method. Anal Biochem 154:255–266.

Tornqvist, M, Fred, C, Haglund, J, Helleberg, H, Paulsson, B, Rydberg, P (2002) Protein adducts: quantitative and qualitative aspects of their formation, analysis and applications. J Chromatogr B Analyt Technol Biomed Life Sci 778:279–308.

Tretyakova, NY, Chiang, SY, Walker, VE, Swenberg, JA (1998) Quantitative analysis of 1,3-butadiene-induced DNA adducts in vivo and in vitro using liquid chromatography electrospray ionization tandem mass spectrometry. J Mass Spectrom 33:363–376.

Trochimowicz, HJ, Loser, E, Feron, VJ, Clary, JJ, Valentine, R (1998) Chronic inhalation toxicity and carcinogenicity studies on b-chloroprene in rats and hamsters. Inha Toxicol 10:443–472.

USEPA (1999) Draft Revised Guidelines for Carcinogen Risk Assessment. http://cfpub.epa.gov/ncea/raf/recordisplay.cfm?deid=54932&CFID=3302774&CFTOKEN=64129803 (3-14-2006).

USEPA (2000) Compendium of methods for the determination of toxic organic compounds in ambient air; Compendium Method TO-15. http://www.epa.gov/ttn/amtic/files/ambient/airtox/to-15r.pdf (12-5-2000).

USEPA (2002) Health Assessment of 1,3-Butadiene. EPA/600/P-98/001F. U.S. Environmental Protection Agency, Office of Research and Development, National Center for Environmental Assessment, Washington Office, Washington, DC.

USEPA and IRIS (2002) 1,3-Butadiene. http://cfpub.epa.gov/iris/quickview.cfm?substance_nmbr=0139 (12-6-2005).

USEPA (2005a) Guidelines for Carcinogen Risk Assessment and Supplemental Guidance for Assessing Susceptibility from Early-Life Exposure to Carcinogens. http://cfpub.epa.gov/ncea/raf/recordisplay.cfm?deid=116283 (12-6-2005a).

USEPA (2005b) Memorandum: Application of New Cancer Guidelines. http://oaspub.epa.gov/eimscomm.getfile?p_download_id=439806 (12-6-2005b).

USEPA (2005c) TRI Explorer–Releases: Chemical Report–1,3-Butadiene. http://www.epa.gov/triexplorer/facility.htm (10-11-2005c).

USEPA (2005d) TRI Explorer–Releases: Chemical Report–Chloroprene. http://www.epa.gov/triexplorer/facility.htm (10-11-2005d).

Valentine, R, Himmelstein, MW (2001) Overview of the acute, subchronic, reproductive, developmental and genetic toxicology of beta-chloroprene. Chem-Biol Interact 135–136:81–100.

van Sittert, NJ, Megens, HJ, Watson, WP, Boogaard, PJ (2000) Biomarkers of exposure to 1,3-butadiene as a basis for cancer risk assessment. Toxicol Sci 56:189–202.

van Steeg, H (2001) The role of nucleotide excision repair and loss of p53 in mutagenesis and carcinogenesis. Toxicol Lett 120:209–219.

Waddell, WJ (2002) Thresholds of carcinogenicity of flavors. Toxicol Sci 68:275–279.

Waddell, WJ (2003a) Comparison of human exposures to selected chemicals with thresholds from NTP carcinogenicity studies in rodents. Hum Exp Toxicol 22:501–506.

Waddell, WJ (2003b) Thresholds in chemical carcinogenesis: what are animal experiments telling us? Toxicol Pathol 31:260–262.

Waddell, WJ (2003c) Thresholds of carcinogenicity in the ED01 study. Toxicol Sci 72:158–163.

Waddell, WJ (2005) Comparisons of thresholds for carcinogenicity on linear and logarithmic dosage scales. Hum Exp Toxicol 24:325–332.

Ward, JB Jr, Ammenheuser, MM, Bechtold, WE, Whorton, EB Jr, Legator, MS (1994) hprt mutant lymphocyte frequencies in workers at a 1,3-butadiene production plant. Environ Health Perspect 102(suppl 9):79–85.
Washington Department of Ecology (2006) Annual data summary, 1,3-butadiene. https://fortress.wa.gov/ecy/aqp/Toxics/1,3-butadiene.shtml (3-10-2006).
Watkinson, R, Somerville, H (1976) The microbial utilization of butadiene. Shell Research Limited, Sittingbourne Research Centre, Kent, UK.
Watson, WP, Cottrell, L, Zhang, D, Golding, BT (2001) Metabolism and molecular toxicology of isoprene. Chem-Biol Interact 135–136:223–238.
Westphal, GA, Blaszkewicz, M, Leutbecher, M, Muller, A, Hallier, E, Bolt, HM (1994) Bacterial mutagenicity of 2-chloro-1,3-butadiene (chloroprene) caused by decomposition products. Arch Toxicol 68:79–84.
Wickliffe, JK, Galbert, LA, Ammenheuser, MM, Herring, SM, Xie, J, Masters, OE III, Friedberg, EC, Lloyd, RS, Ward, JB Jr (2006) 3,4-Epoxy-1-butene, a reactive metabolite of 1,3-butadiene, induces somatic mutations in Xpc-null mice. Environ Mol Mutagen 47:67–70.
Wilson, RH (1944) Health hazards encountered in the manufacture of synthetic rubber. JAMA 124:701.
Wiseman, RW, Cochran, C, Dietrich, W, Lander, ES, Soderkvist, P (1994) Allelotyping of butadiene-induced lung and mammary adenocarcinomas of B6C3F1 mice: frequent losses of heterozygosity in regions homologous to human tumor-suppressor genes. Proc Natl Acad Sci USA 91:3759–3763.
Wistuba, D, Weigand, K, Peter, H (1994) Stereoselectivity of in vitro isoprene metabolism. Chem Res Toxicol 7:336–343.
WJCCTF (2006) West Jefferson County Community Task Force, Air Toxics Monitoring Sites Data. http://208.249.124.184/EJP2/Air_Quality/Database/ (3-10-2006).
Yaws, CL (1976) Physical and thermodynamic properties. Part 18. Butadiene, isoprene and chloroprene. Chem Eng 83:107–115.
Zeiger, E, Anderson, B, Haworth, S, Lawlor, T, Mortelmans, K, Speck, W (1987) *Salmonella* mutagenicity tests: III. Results from the testing of 255 chemicals. Environ Mutagen 9(suppl 9):1–109.
Zhao, C, Vodicka, P, Sraml, RJ, Hemminki, K (2000) Human DNA adducts of 1,3-butadiene, an important environmental carcinogen. Carcinogenesis (Oxf) 21:107–111.
Zhao, C, Vodicka, P, Sram, RJ, Hemminki, K (2001) DNA adducts of 1,3-butadiene in humans: relationships to exposure, GST genotypes, single-strand breaks, and cytogenetic end points. Environ Mol Mutagen 37:226–230.
Zhuang, SM, Cochran, C, Goodrow, T, Wiseman, RW, Soderkvist, P (1997) Genetic alterations of p53 and ras genes in 1,3-butadiene- and 2′,3′-dideoxycytidine-induced lymphomas. Cancer Res 57:2710–2714.
Zhuang, SM, Wiseman, RW, Soderkvist, P (2002) Frequent mutations of the Trp53, Hras1 and beta-catenin (Catnb) genes in 1,3-butadiene-induced mammary adenocarcinomas in B6C3F1 mice. Oncogene 21:5643–5648.

Manuscript received March 5; accepted April 19, 2006.

[illegible]

Index

Algicide, simazine, 1
Arsenic, agricultural sustainability Bangladesh, 78
Arsenic, Bangladesh agriculture, 65
Arsenic, Bangladesh foods, 71, 73
Arsenic, Bangladesh soil & crops, 67
Arsenic behavior, soil, 44
Arsenic chemistry, in crops, 48
Arsenic chemistry, in foods, 60
Arsenic chemistry, in soil, 44
Arsenic chemistry, iron plaque anaerobic soil, 48
Arsenic chemistry, marine plants, 64
Arsenic chemistry, seafoods, 64
Arsenic chemistry, soil rhizosphere, 48
Arsenic, Chinese coal contamination, 92
Arsenic, Chinese coal distribution (map), 93
Arsenic, Chinese coal hazard, 92
Arsenic, epidemic in China, 94
Arsenic, food safety guidelines Bangladesh, 79
Arsenic, grain yield effects, 57
Arsenic, health effects indoor Chinese coal burning, 89 ff.
Arsenic, human exposure Bangladesh, 75
Arsenic, human exposure guidelines, 76
Arsenic, human exposure management, 77
Arsenic, hydroponics effects plants, 55
Arsenic III (AsIII), plant uptake, 50
Arsenic III (AsIII), soil, 45
Arsenic, groundwater major health concern, 43 ff.
Arsenic, irrigation water, 65
Arsenic, irrigation water plant effects, 58
Arsenic, methylated species in plants, 52
Arsenic movement, groundwater & soil to crops, 43 ff.
Arsenic, plant accumulation, 51
Arsenic, plant metabolism, 51
Arsenic, plant tolerance, 54
Arsenic, plant translocation, 51
Arsenic, redox conditions, 45
Arsenic, role of soil iron hydroxides, 44
Arsenic, soil microorganisms effect, 47
Arsenic, soil organic matter effect, 47
Arsenic, soil *pH* effect, 47
Arsenic speciation, foods, 60
Arsenic speciation, rice, 61, 72
Arsenic speciation, soil, 45
Arsenic species, relative toxicity plants, 53
Arsenic, spiked soil tests AsIII/AsV, 57
Arsenic tolerance, plants & iron plaque, 49
Arsenic, toxic effects plants, 52
Arsenic, toxic effects soil microorganisms, 59
Arsenic, uptake in plants, 50
Arsenic V (AsV), phosphate analogue, 47
Arsenic V (AsV), plant uptake, 50
Arsenic V (AsV), soil, 45
Arsenic, volatilization from soil, 48
Arsenobetaine (AsB), in foods, 60
Arthrobacter aurescens, simazine degrader, 15

1,3-**B**utadiene toxicology, 131 ff.
Bangladesh, arsenic situation agriculture, 65
Bangladesh, food & water consumption, 74
Biethylene (butadiene), 132
Bioaccumulation, methylmercury fish, 108

Biomagnification, methylmercury fish, 108
Biomarkers, butadiene exposure, 142
Biomarkers, chloroprene exposure, 155
Biomarkers, human exposure to pollutants, 109
Biomarkers, isoprene exposure, 162
Biomass, alcohol production, 25 ff.
Bivinyl (butadiene), 132
Butadiene, carcinogenicity, 136
Butadiene, chronic inhalation studies, 136
Butadiene, cytochrome P450 oxidation, 137
Butadiene, detoxification mechanisms, 140
Butadiene diepoxide, potent genotoxicity, 139
Butadiene epoxide metabolites, genotoxicity, 139
Butadiene, exposure biomarkers, 142
Butadiene, genotoxicity, 145
Butadiene, hematopoietic toxicity, 135
Butadiene, human epidemiology, 148
Butadiene, industrial uses, 133
Butadiene, kinetics, 137
Butadiene, *Known to be a Human Carcinogen*, 149
Butadiene, metabolic scheme (chart), 138
Butadiene, metabolism, 137, 138
Butadiene, motor vehicle production, 134
Butadiene, multisite carcinogen in mice, 136
Butadiene, mutagenicity, 145
Butadiene, number workers exposed, 133
Butadiene, occupational exposure, 133
Butadiene, other chemical names, 132
Butadiene, photodegradation rates, 134
Butadiene, physicochemical properties, 132
Butadiene, *Probably Carcinogenic to Humans*, 149
Butadiene, risk assessment, 150
Butadiene, toxicity, 134
Butadiene toxicology, 131 ff.
Butadiene, worker exposure leukemia, 149
Butadiene, world production, 133
Butadiene-1,3 (butadiene), 132

China, largest global coal producer/consumer, 90
Chinese coal, arsenic hazard, 92
Chinese coal, indoor burning health effects, 89 ff.
Chloroprene, exposure biomarkers, 155
Chloroprene, genotoxicity, 155
Chloroprene, human epidemiology, 156
Chloroprene, industrial uses, 150
Chloroprene, kinetics, 152
Chloroprene, metabolic scheme (chart), 154
Chloroprene, metabolism, 152, 154
Chloroprene, physical properties, 133
Chloroprene, physicochemical properties, 150
Chloroprene, risk assessment, 157
Chloroprene, toxicity, 151
Chloroprene toxicology, 131 ff.
Chloroprene, worker exposure, 151
Chloroprene, world production, 150
CIMMYT-Bangladesh, arsenic research, 70
Coal burning, indoors without chimneys, 91
Coal burning, trace element dispersal, 90
Corn ethanol, input costs, 28
Corn fermentation/distillation, energy inputs, 28
Corn production, energy inputs, 26, 27
Corn use in ethanol production, 26
Cytochrome P-450 oxidation, simazine, 17

Degradation pathways, simazine, 8
Demethylation, methylmercury in organisms, 108
Dental fluorosis, case photos China, 97
Dimethylarsenic acid (DMA), in soil, 45
Divinyl (butadiene), 132
DMA (dimethylarsenic acid), in soil, 45

Earthworms, simazine toxicity, 2
Energy inputs, corn production, 26
Energy inputs, fermentation/distillation, 28
Erythrene (butadiene), 132
Ethanol, annual production, 30
Ethanol as gasoline additive, 25 ff.
Ethanol production, for vehicle fuel, 25 ff.
Ethanol production, by-products, 30
Ethanol production, cornland use, 30
Ethanol production, cropland use, 31
Ethanol production, economic costs, 29
Ethanol production, energy costs, 25 ff.
Ethanol production, energy return, 33
Ethanol production, environmental impacts, 32
Ethanol production, federal subsidies, 29
Ethanol production, food security, 34
Ethanol production, food vs fuel issue, 35
Ethanol production, net energy yield, 29
Ethanol production, sugarcane, 31

Federal subsidies, ethanol production, 29
Fermentation/distillation, energy inputs, 28
Fish consumption rate, versus hair mercury content, 107 ff.
Fish, methylmercury biomagnification, 108
Fluorine, Chinese coal hazard, 96
Fluorine, dental fluorosis in China, 97
Fluorine, epidemic in China, 96
Fluorine health effects, indoor Chinese coal burning, 89 ff.

Gasohol production, 25 ff.
Genotoxicity, butadiene, 145
Genotoxicity, chloroprene, 155
Genotoxicity, isoprene, 162
Groundwater contamination, simazine, 5, 12

Hair, bioindicator of mercury exposure, 107 ff.
Hair, biomarker of mercury exposure, 109
Hair, growth rate, 109
Hair mercury content, versus fish consumption rate, 107 ff.
Hair mercury, correlated with blood mercury, 108
Hair-mercury, different populations, 114
Herbicide abiotic degradation, simazine, 14
Herbicide biodegradation, simazine, 15
Herbicide management practices, simazine, 10
Herbicide soil half-lives, simazine, 3
Herbicides, groundwater contamination, 5
Human brain damage, methylmercury, 108
Human epidemiology, butadiene, 148
Human epidemiology, chloroprene, 156
Human epidemiology, isoprene, 164
Hydroponics, arsenic effects plants, 55

Indoor coal burning, health effects China, 89 ff.
Iron hydroxides, role in soil arsenic, 45
Iron plaque, role in plant arsenic tolerance, 49
Iron plaque, roots wetland plants, 49
Isoprene, exposure biomarkers, 162
Isoprene, genotoxicity, 162
Isoprene, human epidemiology, 164
Isoprene, industrial uses, 158
Isoprene, kinetics, 160
Isoprene, metabolic scheme (chart), 160
Isoprene, metabolism, 160, 161
Isoprene, physical properties, 133
Isoprene, physicochemical properties, 158
Isoprene, toxicity, 159
Isoprene toxicology, 131 ff.
Isoprene, world production, 158

Leukemia, butadiene worker exposure, 149
Lymnea stagnalis (snail), simazine toxicity, 2

Malnourished world population percentage, 35
Mercury, daily human exposure estimates, 112
Mercury exposure, hair as bioindicator, 107 ff.
Mercury fungicides, poisonings in Iraq, 121
Mercury, hair content versus fish consumption, 107 ff.
Mercury, hair levels Alaska, 118
Mercury, hair levels Amazonia, 116
Mercury, hair levels Arabia, 119
Mercury, hair levels Asian populations, 115
Mercury, hair levels Cambodia, 120
Mercury, hair levels correlate with blood levels, 108
Mercury, hair levels from different populations, 114
Mercury, hair levels Spain, 119
Mercury, hair levels Sweden, 117
Mercury, hair levels Tanzania, 120
Mercury, human exposure case studies, 113
Mercury, human exposure estimates, 112
Mercury, human exposure workplaces, 110
Mercury, inorganic history, 110
Mercury, poisonings in Iraq, 121
Mercury vapor, binding to hair, 109
Mercury vapor, long range transport, 110
Methylmercury, bioaccumulation fish, 108
Methylmercury, biomagnification fish, 108
Methylmercury, blood vs hair levels, 113
Methylmercury, demethylation in organisms, 108
Methylmercury, hair follicle incorporated, 109
Methylmercury, hair samples, 109
Methylmercury, history, 110
Methylmercury, human brain damage, 108
Methylmercury, Minamata Disease Japan, 113
Methylmercury, naturally occurring compound, 107
Minamata Disease in Japan, methylmercury poisoning, 113
MMA (monomethylarsenic acid), in soil, 45
Mode of action, simazine, 4
Monomethylarsenic acid (MMA), in soil, 45
Mutagenicity, butadiene, 145

Organomercurial fungicides, poisonings in Iraq, 121

PBPK (physiologically based pharmacokinetic) models, 141
Penicillium steckii, simazine degrading, 16
Pesticide management practices, simazine, 10
Phosphate, analogue of Arsenic V, 47
Photolysis, simazine, 15
Physicochemical properties, simazine, 2
Physiologically Based Pharmacokinetic (PBPK) Models, 141
Plant tolerance, arsenic, 54

Redox conditions, arsenic soil role, 45

Selenium, Chinese coal contamination, 98
Selenium, Chinese coal hazard, 98
Selenium, epidemic in China, 98
Selenium health effects, indoor Chinese coal burning, 89 ff.
Selenium, species in nature, 98
Selenosis, case photos China, 99
Simazine, abiotic degradation, 14
Simazine, agricultural management practices, 10
Simazine, air behavior, 13
Simazine, algicide, 1
Simazine, annual use volume, 1
Simazine, available formulations, 2
Simazine, biodegradation, 15
Simazine, chemical name, 1
Simazine, chemical structure, 3
Simazine, chemical synthesis method, 4
Simazine, chemodynamics, 4

Simazine, crustacean toxicity, 1
Simazine, degradation pathways, 8
Simazine, earthworm toxicity, 2
Simazine, groundwater contamination California, 5
Simazine herbicide, chemistry and fate, 1 ff.
Simazine, hydrolysis, 14
Simazine, in rainfall, 13
Simazine metabolism, cytochrome P-450 oxidation, 17
Simazine, mode of action, 4
Simazine, orchard management, 10
Simazine, oxidation, 14
Simazine, pesticide management practices, 10
Simazine, photolysis, 15
Simazine, physicochemical properties, 2
Simazine, precipitation, 13
Simazine, proprietary names, 2
Simazine, snail toxicity, 2
Simazine, soil behavior, 4
Simazine, soil half-lives, 3
Simazine, soil runoff characteristics, 9
Simazine, soil sorption, 4, 6
Simazine, sorption to organic matter, 7
Simazine, surface water contamination, 12
Simazine, toxicity to crustacean, 1
Simazine, toxicity to earthworms, 2
Simazine, water behavior, 11
Simazine, water quality criteria, 13
Snails, simazine toxicity, 2
Soil microorganisms, arsenic toxicity, 59
Soil microorganisms, affect arsenic availability, 47
Soil organic matter, affect arsenic availability, 47
Soil *pH*, affect arsenic availability, 47
Sugarcane, ethanol production, 31
Surface water contamination, simazine, 12

Trace element dispersal, coal burning, 90
Trace elements, coal, 91
Triazine herbicides, 2

Vinylethylene (butadiene), 132

Water quality criteria, simazine, 13

MIX
Papier aus verantwortungsvollen Quellen
Paper from responsible sources
FSC® C105338
FSC
www.fsc.org

If you have any concerns about our products,
you can contact us on
ProductSafety@springernature.com

In case Publisher is established outside the EU,
the EU authorized representative is:
Springer Nature Customer Service Center GmbH
Europaplatz 3, 69115 Heidelberg, Germany

Printed by Libri Plureos GmbH
in Hamburg, Germany